SpringerBriefs in Water Science and Technology

SpringerBriefs in Water Science and Technology present concise summaries of cutting-edge research and practical applications. The series focuses on interdisciplinary research bridging between science, engineering applications and management aspects of water. Featuring compact volumes of 50 to 125 pages (approx. 20,000–70,000 words), the series covers a wide range of content from professional to academic such as:

- Literature reviews
- In-depth case studies
- Bridges between new research results
- Snapshots of hot and/or emerging topics

Topics covered are for example the movement, distribution and quality of freshwater; water resources; the quality and pollution of water and its influence on health; and the water industry including drinking water, wastewater, and desalination services and technologies.

Both solicited and unsolicited manuscripts are considered for publication in this series.

Manish Kumar · Durga Prasad Panday ·
Vivek Agarwal

Bridging Water Quality and Quantity in the Anthropocene

Climate Change and Sustainability

Springer

Manish Kumar
Escuela de Ingeniería y Ciencias
Tecnologico de Monterrey
Monterrey, Nuevo León, Mexico

Sustainability Cluster, School of Advanced Engineering
UPES
Dehradun, Uttarakhand, India

Durga Prasad Panday
Escuela de Ingeniería y Ciencias
Tecnologico de Monterrey
Monterrey, Nuevo León, Mexico

Sustainability Cluster, School of Advanced Engineering
UPES
Dehradun, Uttarakhand, India

Vivek Agarwal
Science and Environment
Northumbria University
Newcastle, UK

ISSN 2194-7244 ISSN 2194-7252 (electronic)
SpringerBriefs in Water Science and Technology
ISBN 978-3-032-25654-6 ISBN 978-3-032-25655-3 (eBook)
https://doi.org/10.1007/978-3-032-25655-3

This Springer imprint is published by the registered company Springer Nature Switzerland AG
The registered company address is: Gewerbestrasse 11, 6330 Cham, Switzerland

Foreword

Water is often discussed as an environmental resource, an engineering challenge, or a public policy priority. Yet, from the perspective of food security, water is something even more fundamental: it is the invisible foundation of every food system. It sustains agricultural production, safeguards ecosystems, enables food processing, supports nutrition, and ultimately determines whether societies can provide safe, sufficient, and sustainable food for their populations.

For this reason, "Bridging Water Quality and Quantity in the Anthropocene: Climate Change and Sustainability" is both timely and necessary. The central message of this book is clear: water quantity and water quality can no longer be understood, modeled, or governed as separate dimensions. In a changing climate, the availability of water is inseparable from its safety, usability, and long-term sustainability. Scarcity is not only a matter of volume; it is also a matter of quality. Water that is contaminated, saline, biologically unsafe, or unsuitable for irrigation represents a form of functional scarcity, even when it is physically present.

This perspective is especially relevant for those of us working in food security. Agriculture remains one of the largest users of freshwater globally, and the stability of food production depends on reliable access to water in both adequate quantity and acceptable quality. Droughts reduce irrigation capacity and crop yields, while floods can mobilize contaminants, nutrients, sediments, and pathogens into water bodies and agricultural landscapes. Groundwater depletion threatens long-term productivity, and the deterioration of water quality can compromise soils, crops, livestock, fisheries, and human health. In this sense, water insecurity rapidly becomes food insecurity.

The Anthropocene has intensified these interdependencies. Climate variability, urbanization, industrial activity, agricultural intensification, land-use change, and emerging contaminants are reshaping hydrological systems in ways that challenge traditional assumptions. Historical patterns are no longer sufficient to guide future decisions. The combined pressures of climate change and human activity require integrated approaches that can anticipate risk, support adaptive governance, and protect both ecosystems and communities.

One of the strengths of this book is that it recognizes water systems as coupled socio-environmental systems. The authors move beyond a narrow technical treatment of water management and instead examine the interactions among hydrology, contamination, climate extremes, modelling, governance, carbon neutrality, and human well-being. This integrated framing is essential for the future of food systems. Ensuring food security in the twenty-first century will require not only producing more food, but producing it in ways that preserve water resources, reduce environmental burdens, and strengthen resilience under uncertainty.

The book also highlights the growing importance of predictive modelling, artificial intelligence, remote sensing, and decision-support systems for water management. These tools are becoming increasingly relevant for anticipating contamination events, identifying vulnerable aquifers, understanding drought-related quality changes, and guiding timely interventions. For food security, such approaches can help inform irrigation strategies, protect agricultural watersheds, reduce exposure to unsafe water, and support more resilient production systems. However, the value of these technologies depends on their integration with governance, policy, and local realities. Prediction must ultimately translate into prevention, adaptation, and action.

Equally important is the book's emphasis on carbon-neutral water resilience. Many conventional responses to water scarcity, such as deep groundwater pumping, long-distance transfers, desalination, and advanced treatment, can be energy-intensive. As climate change accelerates, water solutions must avoid reinforcing the emissions pathways that deepen future water risks. Nature-based solutions, efficient irrigation, wastewater reuse, smart monitoring, renewable energy integration, and adaptive management offer opportunities to strengthen both water and food systems while contributing to broader sustainability goals.

This volume therefore speaks to researchers, engineers, policymakers, and practitioners across disciplines. It reminds us that water sustainability is not a sectoral issue, but a foundation for health, food security, climate resilience, economic development, and social equity. The future of sustainable food systems will depend on our ability to manage water not only as a resource to be allocated, but as a living system whose quality and quantity are deeply interconnected.

I am pleased to contribute this foreword from the perspective of food security and applied science. The themes developed in this book are directly aligned with the urgent need to build resilient, sustainable, and equitable systems capable of nourishing a growing population under increasingly complex environmental conditions. By bridging water quality and quantity, the authors provide a valuable framework for understanding one of the defining challenges of our time.

Daniel A. Jacobo-Velázquez, Ph.D.
Associate Dean of Research
and Scientific Graduate Programs
School of Engineering and Sciences,
Tecnológico de Monterrey
Monterrey, Mexico

Preface

The growing complexity of global water challenges has made it increasingly important to understand the interconnected nature of water quantity, water quality, climate change, sustainability, and environmental resilience. Rapid urbanization, changing climate regimes, groundwater depletion, pollution, and ecosystem degradation are collectively reshaping water systems across the world. In this context, "***Bridging Water Quality and Quantity in the Anthropocene: Climate Change and Sustainability***" was developed to provide an integrated scientific perspective on these evolving challenges and their broader sustainability implications.

This book brings together discussions on climate and anthropogenic impacts on water systems, the interlinkages between water quality and quantity, predictive contaminant modelling, resilience strategies, and carbon-neutral approaches for sustainable water management. Through these chapters, we aim to highlight the need for interdisciplinary thinking and adaptive solutions capable of addressing the uncertainties associated with the Anthropocene.

The preparation of this book was made possible through the encouragement and support of institutions committed to sustainability, innovation, and interdisciplinary research. We sincerely acknowledge Tecnológico de Monterrey, Mexico, for fostering global scientific collaboration and promoting research in sustainability, climate resilience, and environmental systems. We are equally grateful to UPES, India, for encouraging interdisciplinary research and academic initiatives related to sustainability, environmental engineering, and water resource management. We also acknowledge Northumbria University, United Kingdom, for supporting research collaboration and advancing scientific efforts in environmental sustainability and water-related studies.

We would also like to express our heartfelt appreciation to Scholar Sachin Tripathi, Shiwangi Dogra, Aanchal Kumari, and Kanika Dogra for their valuable support and assistance during the preparation of this book. Their contributions and dedication are sincerely acknowledged.

Finally, we extend our gratitude to researchers, collaborators, students, and the wider scientific community whose continued efforts inspire meaningful advancements in water science and sustainability research. We hope this book will serve as a useful resource for researchers, students, policymakers, and professionals working toward resilient and sustainable water systems in a changing world.

Monterrey, Mexico
Dehradun, India
Newcastle, UK

Manish Kumar
Durga Prasad Panday
Vivek Agarwal

Contents

1 Climate and Anthropogenic Impacts on Water Systems and Sustainability 1
1.1 Introduction 2
1.2 Climate Change and Water Systems 3
1.2.1 Altered Climate Regimes and Hydrological Variability 3
1.2.2 Cryospheric Decline and Impact on Water Systems 4
1.3 Anthropogenic Impacts on Water Systems 5
1.3.1 Urbanization and Hydrological Modification 5
1.3.2 Anthropogenic Water Use and Its Implications 9
1.3.3 Land-Use Change and Soil–Water Interactions 9
1.4 Combined Effects of Climate and Human Activities 10
1.5 Implications for Water Sustainability 12
1.5.1 Balancing Supply, Demand, and Ecological Thresholds 12
1.5.2 Data Gaps, Uncertainty, and Socioeconomic Vulnerability 13
1.6 Governance Transformation and Technological Innovation 13
1.7 Conclusions 16
References 18

2 Interlinkages Amongst Water Quality and Quantity, Humanity, and Sustainability 27
2.1 Introduction 29
2.2 Trends and Evolution in Water Quantity and Quality Modelling 31
2.2.1 Evolution of Water Quality Models 31
2.2.2 Evolution of Water Quantity Models 31
2.2.3 Integration of Climate Indices into Water Quality–Quantity Studies 32

2.3 Humanity and Sustainability Dimensions of Water Quality–Quantity Linkages 34
2.3.1 The Human–Water Nexus: Health, Food Security, and Livelihoods 39
2.3.2 Water Quality–Quantity Linkages in Sustainability Frameworks 40
2.3.3 Integrating Humanity and Sustainability into Modelling Frameworks 41
2.4 Future Directions 43
2.5 Conclusion 44
References 45

3 Predictive Modelling for Contaminant Forecasting 49
3.1 Introduction 50
3.2 Theoretical Framework for Contaminant Transport 51
3.2.1 Fundamental Transport Processes 51
3.2.2 Mathematical Foundations 53
3.3 Predictive Modelling Approaches 56
3.3.1 Data-Driven Models 56
3.3.2 Hybrid Approaches 57
3.3.3 Empirical Models 58
3.3.4 Process-Based Models 58
3.4 Model Development and Implementation 59
3.4.1 Data Requirements and Preprocessing 59
3.4.2 Model Calibration and Validation 64
3.4.3 Uncertainty Analysis 66
3.5 Integration with Decision Support Systems 67
3.5.1 Early Warning Systems and Risk Assessment 67
3.5.2 Operational Decision Support Tools 69
3.5.3 Stakeholder Communication and Policy Support 69
3.6 Future Directions and Challenges 70
3.6.1 Artificial Intelligence and Big Data Integration 70
3.6.2 Climate Change and Emerging Contaminants 71
3.6.3 Digital Twins and Research Priorities 72
3.7 Conclusions 72
References 73

4 Enhancing the Water Resilience with Special Emphasis on Carbon Neutrality 77
4.1 Introduction: The Water-Carbon Nexus 78
4.2 Understanding Water Resilience 82
4.2.1 Components of a Resilient Water System 82
4.2.2 Threats to Water Resilience 83
4.2.3 Measuring and Benchmarking Water Resilience 84

4.3 The Carbon Footprint of Water Systems ... 85
4.3.1 Energy Consumption in Water Extraction, Treatment, and Distribution ... 86
4.3.2 Emissions from Wastewater Management ... 88
4.3.3 Embodied Carbon in Water Infrastructure ... 89
4.4 Strategies and Innovations for Carbon–Neutral Water Resilience ... 92
4.4.1 Energy Efficiency and Demand Management ... 92
4.4.2 Renewable Energy Integration and Energy Recovery ... 94
4.4.3 Nature-Based Solutions and Green Infrastructure ... 94
4.4.4 Smart Systems and Advanced Technologies ... 95
4.5 Policy Frameworks, Governance, and Overcoming Barriers ... 96
4.5.1 Integrating Water and Climate Policies ... 96
4.5.2 Financing Carbon–Neutral Water Infrastructure ... 97
4.5.3 Technical, Financial, and Institutional Barriers ... 97
4.5.4 Strategies for Accelerating Progress ... 98
4.6 Conclusion ... 99
References ... 100

5 Summary and Future Directions ... 103
5.1 Integrated Synthesis of Climate, Anthropogenic Pressures, and Water System Sustainability ... 103
5.2 Interlinkages Amongst Water Quality and Quantity—Humanity and Sustainability ... 104
5.2.1 Water Quality Constraints as Effective Water Scarcity ... 106
5.2.2 Human Health and Exposure Pathways in Coupled Systems ... 107
5.2.3 Climate Variability, Drought Indices, and Quantity–Quality Coupling ... 107
5.2.4 Governance, Uncertainty, and Integrated Decision-Making ... 108
5.3 Predictive Modelling for Contaminant Forecasting ... 109
5.3.1 Data-Driven Approaches and the Role of Artificial Intelligence ... 109
5.3.2 Climate Indices and Teleconnections as Predictors in Contamination Forecasting ... 110
5.3.3 Model Evaluation, Uncertainty, and Transferability ... 111
5.4 Enhancing Water Resilience with Special Emphasis on Carbon Neutrality ... 111
5.4.1 Nature-Based and Hybrid Interventions as Low-Carbon Resilience Pathways ... 112
5.4.2 The Water–Energy–Carbon Nexus in Supply and Treatment Choices ... 112
5.4.3 Climate-Informed Operation and Adaptive Resilience ... 113

5.5 Roadmap for Next-Generation Water Sustainability 114
5.6 Conclusions 115
References 118

Chapter 1
Climate and Anthropogenic Impacts on Water Systems and Sustainability

Abstract Globally, water systems are increasingly influenced by the combined pressures of climate change and anthropogenic activities, posing significant challenges to water security and long-term sustainability. Alterations in precipitation patterns, rising temperatures, and the increasing frequency of extreme events are reshaping hydrological processes, while rapid urbanization, agricultural expansion, industrial development, and escalating pollution continue to degrade water quality and ecological integrity. These interacting drivers exert compounded pressures on surface water and groundwater resources, aquatic ecosystems, and the socioeconomic systems that depend on them. This chapter examines the impacts of climate change and human activities on water systems, with particular attention to changes in precipitation regimes, hydrological cycles, river systems, groundwater recharge and depletion, and environmental health. It highlights the interconnected influences of urbanization, agriculture, and pollution, and how these, together with climatic variability, affect water availability, water quality, biodiversity, public health, food security, and socioeconomic development. The chapter further considers water sustainability as a dynamic socio-environmental challenge shaped by governance structures, institutional capacity, information availability, and uncertainty, alongside technical and management approaches such as Integrated Water Resources Management, climate-resilient infrastructure, nature-based solutions, and stakeholder engagement. Finally, it identifies future research priorities and emphasizes the importance of interdisciplinary and transdisciplinary approaches to enhance the resilience of water systems and support their sustainable management under rapidly changing environmental conditions.

Keywords Climate change · Water sustainability · Surface water-groundwater interactions · Anthropogenic impacts · Integrated water resources management

M. Kumar et al., *Bridging Water Quality and Quantity in the Anthropocene*, SpringerBriefs in Water Science and Technology,
https://doi.org/10.1007/978-3-032-25655-3_1

1.1 Introduction

Water systems underpin ecological stability, economic productivity, and human well-being. Surface water and groundwater collectively sustain drinking water supply, agriculture, industry, energy production, and ecosystem functioning. Their long-term viability depends on maintaining both quantitative reliability and qualitative integrity under increasing environmental stress. Over recent decades, water systems worldwide have experienced intensifying pressures driven by climate change and expanding anthropogenic activities (Goswami and Bisht 2017; Agarwal et al. 2024).

Climate change is altering the hydrological cycle through modifications in precipitation regimes, temperature patterns, and the frequency of extreme events. Changes in evaporation rates, atmospheric moisture content, and rainfall distribution are reshaping spatial and temporal water availability (Kumar et al. 2026; Agarwal et al. 2025). The growing intensity of floods, droughts, and heatwaves introduces hydrological non-stationarity, reducing the reliability of historical records as a basis for planning (Allan et al. 2020). Accelerated glacier and snowmelt modify seasonal runoff patterns and, in some regions, increase vulnerability to saline intrusion in coastal aquifers, affecting freshwater reserves (Asif et al. 2023; Kumar et al. 2023; Panday and Kumar 2024).

Anthropogenic pressures compound these climatic influences. Rapid urbanization, industrial expansion, and agricultural intensification have altered natural hydrological pathways, increased water abstraction, and amplified pollutant loads. Excessive withdrawal from rivers and aquifers disrupts baseflow dynamics and groundwater recharge processes, while untreated or partially treated wastewater introduces nutrients, heavy metals, pharmaceuticals, and other emerging contaminants into aquatic systems (Akhtar et al. 2021; Gavrilaş et al. 2025). Agricultural runoff enriched with fertilizers and pesticides contributes to eutrophication, hypoxia, and long-term ecosystem degradation (Häder et al. 2020; Akhtar et al. 2021). Land-use transformations, including deforestation, river regulation, and channel modification, further reduce hydrological buffering capacity and ecological resilience (Gavrilaş et al. 2025; Panday et al. 2025).

The interaction between climate variability and human interventions generates compounded effects on water systems. Reduced river discharge during drought conditions elevates pollutant concentrations through diminished dilution capacity, whereas intense rainfall events mobilize accumulated contaminants from urban and agricultural landscapes into receiving water bodies (Muruganandam et al. 2023). Increased groundwater abstraction under prolonged dry periods accelerates aquifer depletion and land subsidence, undermining long-term water security (Mahdian et al. 2023; Muruganandam et al. 2023). These interconnected processes influence not only hydrological balances but also the chemical and ecological state of water resources.

Consequences extend beyond hydrology into food systems, public health, energy production, and economic development. Water insecurity constrains agricultural productivity, increases exposure to waterborne diseases, and heightens socioeconomic vulnerability, particularly among marginalized populations (Brewis et al.

2020; Tariq et al. 2024). Degradation of aquatic ecosystems reduces biodiversity and weakens ecosystem services that regulate nutrient cycling, habitat provision, and climate feedback mechanisms.

Addressing these pressures requires integrated approaches that recognize the interdependence of climatic drivers and anthropogenic influences. Sustainable water management must reconcile environmental protection with social and economic demands in the face of uncertainty. Climate-resilient infrastructure, pollution prevention, ecosystem-based solutions, and coordinated governance structures are essential components of adaptive water systems (Pahl-Wostl 2020; Lee et al. 2022; Panday et al. 2024; Lindner et al. 2025). This chapter examines the combined influence of climate change and human activities on water systems, focusing on hydrological alterations, water quality degradation, ecosystem responses, and sustainability challenges. It establishes the scientific and institutional context within which subsequent chapters explore interlinkages, predictive modelling, and resilience strategies.

1.2 Climate Change and Water Systems

Climate change is reshaping hydrological processes through modifications in atmospheric dynamics, thermal regimes, cryospheric storage, and sea levels. These drivers alter the magnitude, timing, and spatial distribution of water availability while simultaneously influencing water quality, ecosystem stability, and infrastructure performance. The cumulative effect is a transition from historically stationary hydrological conditions toward increasingly variable and uncertain regimes (Allan et al. 2020; Ciampittiello et al. 2024).

1.2.1 Altered Climate Regimes and Hydrological Variability

Changes in precipitation patterns are among the most direct climate-driven influences on water systems. Variations in rainfall intensity, frequency, and seasonality disrupt established hydrological cycles and challenge infrastructure designed under assumptions of long-term climatic stability (Leveque et al. 2021; Palmer et al. 2023). Many regions experience increased inter-annual variability, delayed or advanced monsoon onset, shortened wet seasons, or prolonged dry spells. These shifts affect soil moisture dynamics, surface runoff generation, and aquifer recharge processes (Thomas and Joseph 2024; Wu et al. 2024b). Intensification of rainfall events increases peak flows and flash flood risks, while extended dry periods reduce baseflows and storage replenishment. Heavy precipitation enhances erosion and sediment transport, mobilizing nutrients and contaminants into rivers and reservoirs (Tabari 2020; Zou et al. 2023). Conversely, reduced precipitation and prolonged drought lower dilution capacity in receiving waters, elevating pollutant concentrations and degrading water quality (Mosley 2015; Lu et al. 2025a). These dual effects modify both the quantity and chemical condition of water resources, increasing hydrological instability across basins.

Increasing air and surface temperatures accelerate evaporation and evapotranspiration. Enhanced vapor pressure deficits intensify water losses from rivers, lakes, reservoirs, and soils, particularly in arid and semiarid environments (Althoff et al. 2020; Ajjur and Al-Ghamdi 2021). Elevated evapotranspiration increases irrigation demand and reduces effective recharge, thereby constraining groundwater replenishment (Sadok et al. 2021; Berghuijs et al. 2024). Thermal alterations also affect in-stream processes and lake stratification patterns. Higher water temperatures decrease dissolved oxygen solubility and modify biogeochemical cycling, potentially increasing internal nutrient loading and contaminant mobilization (Wang and Liu 2023). Reduced summer flows combined with elevated temperatures further intensify ecological stress, especially in systems already experiencing abstraction pressures (Tabari 2020).

1.2.2 Cryospheric Decline and Impact on Water Systems

Cryospheric components such as glaciers and seasonal snowpacks regulate freshwater storage and release in many river basins. Rising temperatures reduce snowfall fractions and accelerate snowmelt timing, resulting in earlier spring peak flows and diminished late-summer discharge (Gottlieb and Mankin 2024). Glacier retreat initially increases meltwater contributions but ultimately reduces long-term water availability as ice reserves diminish (Nie et al. 2021). The loss of cryospheric buffering capacity alters seasonal hydrological balance and increases exposure to both water scarcity and extreme events. Reduced summer flows affect irrigation supply, hydropower production, and ecological stability, while destabilized glacial systems increase risks of glacial lake outburst floods (Milner et al. 2017; Nie et al. 2021). These transitions fundamentally modify water storage dynamics in mountain-fed and downstream systems.

Surface water bodies respond rapidly to climatic variability. River flow regimes shift in magnitude and timing due to changes in precipitation, temperature, and upstream cryospheric contributions. Increased rainfall intensity produces higher peak discharges and sediment loads, whereas prolonged dry conditions reduce baseflows and increase flow intermittency (Deng et al. 2025; Wang and Liu 2023). Such alterations disrupt reservoir management, irrigation scheduling, and the maintenance of ecological flows. Lake systems are similarly sensitive. Elevated temperatures and increased evaporation reduce lake levels, particularly in semiarid and closed basins (Jiménez-Bonilla et al. 2025). Fluctuations in water levels influence shoreline stability, habitat distribution, and water quality by concentrating dissolved constituents (Woolway et al. 2022; Soranno et al. 2025). In some basins, declining inflows combined with enhanced evaporation lead to long-term shrinkage and ecological transformation.

Groundwater systems provide temporal buffering against climatic variability, yet changes increasingly influence them in recharge dynamics. High-intensity rainfall events may generate rapid runoff rather than infiltration, reducing effective recharge despite short-term increases in precipitation (Kuang et al. 2024). Rising temperatures and extended dry periods further limit soil moisture availability and percolation, decreasing aquifer replenishment rates (Berghuijs et al. 2024). During droughts, dependence on groundwater abstraction intensifies, often exceeding natural recharge and accelerating depletion (Amanambu et al. 2020). Declining water tables reduce well yields, increase pumping energy requirements, and heighten vulnerability to salinization in coastal aquifers (Dao et al. 2024). Aquifer responses vary by hydrogeological setting: shallow unconfined systems respond rapidly to climatic variability, whereas deep confined aquifers exhibit delayed but potentially irreversible depletion patterns (Barbieri et al. 2023). Table 1.1 illustrates how climate change drivers, ranging from shifting precipitation to cryospheric decline, fundamentally disrupt hydrological processes and the quality–quantity balance of water systems. These disruptions necessitate integrated adaptive management strategies, such as managed aquifer recharge and resilient storage design, to ensure long-term water sustainability.

1.3 Anthropogenic Impacts on Water Systems

Human activities have fundamentally altered hydrological processes, storage dynamics, and water quality pathways. Urban expansion, industrial production, agricultural intensification, land-use transformation, and pollutant discharge collectively modify both water availability and hydrochemical regimes. These pressures operate at basin, watershed, and aquifer scales, frequently amplifying climate-driven variability (Wu et al. 2024a).

1.3.1 Urbanization and Hydrological Modification

Urbanization transforms natural landscapes into impervious surfaces, reducing infiltration and increasing rapid surface runoff. This alteration disrupts groundwater recharge, elevates peak discharge during storm events, and weakens the natural buffering capacity of watersheds (McGrane 2016). Urban drainage systems, historically designed under stationary rainfall assumptions, often lack the capacity to manage intensified precipitation, resulting in recurrent flooding and contaminant mobilization (Pochodyła-Ducka et al. 2023). Stormwater runoff from roads, rooftops, and industrial areas transports sediments, nutrients, hydrocarbons, heavy metals, and microplastics into receiving water bodies. The reduction in vegetated cover further limits natural filtration and evapotranspiration. Sustainable urban drainage systems, including permeable pavements, green infrastructure, rainwater harvesting, and constructed wetlands, have demonstrated potential to enhance infiltration and

Table 1.1 Climate change drivers and impacts on surface water and groundwater systems (Panday and Kumar 2022; Dixit et al. 2022; Asif et al. 2023; Ciampittiello et al. 2024; Allan et al. 2020; Tabari 2020; Eekhout et al. 2018; Mosley 2015; Amanambu et al. 2020; Kuang et al. 2024; Berghuijs et al. 2024; Gottlieb and Mankin 2024; Nie et al. 2021; Milner et al. 2017; Deng et al. 2025; Wang and Liu 2023; Jiménez-Bonilla et al. 2025; Woolway et al. 2022; Soranno et al. 2025; Dao et al. 2024; Barbieri et al. 2023)

Climate change driver	Key hydrological processes affected	Impacts on surface water systems	Impacts on groundwater systems	Water quality implications	Quality–quantity interaction mechanism	Implications for water sustainability	Adaptive management relevance
Changes in rainfall intensity and seasonality	Precipitation distribution, runoff generation, infiltration	Increased flood peaks, reduced baseflows, and altered seasonal flows	Variable recharge rates; reduced infiltration during high-intensity events	Increased contaminant mobilization during floods; reduced dilution during drought	Dilution–concentration shifts driven by discharge variability	Reduced reliability of supply; increased contamination risk	Flexible reservoir operation; managed aquifer recharge (MAR); floodwater capture
Increased frequency of floods and droughts	Soil moisture deficit, rapid runoff, evapotranspiration extremes	Flood-induced sediment and nutrient loading; drought-driven low flows	Increased abstraction during drought; episodic recharge pulses	Nutrient spikes, hypoxia, pollutant concentration under low flow	Extreme-event amplification of contaminant transport	Infrastructure stress: higher treatment costs	Early warning systems; adaptive abstraction controls
Rising air and water temperatures	Evaporation, evapotranspiration, thermal stratification	Reduced storage due to evaporation; altered thermal regimes	Reduced recharge via soil moisture loss; increased salinization risk	Decreased dissolved oxygen; enhanced biogeochemical reactions	Temperature-driven solubility and redox changes	Intensified scarcity and ecological stress	Evaporation control measures; climate-resilient storage design

(continued)

Table 1.1 (continued)

Climate change driver	Key hydrological processes affected	Impacts on surface water systems	Impacts on groundwater systems	Water quality implications	Quality–quantity interaction mechanism	Implications for water sustainability	Adaptive management relevance
Cryospheric decline (glacier and snowpack loss)	Snow accumulation/melt timing; seasonal flow redistribution	Earlier spring peak flows; reduced late-summer discharge	Long-term reduction in recharge in glacier-fed basins	Concentration of dissolved solids during low-flow seasons	Loss of seasonal buffering affects both flow and quality	Reduced seasonal reliability; higher vulnerability downstream	Seasonal flow forecasting; alternative storage strategies
Sea-level rise	Hydraulic gradient alteration; saline intrusion	Salinization of estuaries and coastal surface waters	Saltwater intrusion into coastal aquifers	Increased salinity; mobilization of trace metals	Density-driven intrusion altering groundwater chemistry	Loss of potable freshwater resources	Barrier recharge systems; controlled pumping regimes
Extreme heatwaves	Surface warming; evaporation surges	Rapid reservoir drawdown; algal bloom proliferation	Enhanced evapotranspiration reduces recharge	Harmful algal blooms; toxin release; oxygen depletion	Heat–nutrient synergy intensifying eutrophication	Ecosystem instability; increased treatment demand	Shade restoration; nutrient management; cooling strategies
Compound flood–drought sequences	Alternating storage surpluses and deficits	Sediment flushing followed by low-flow stagnation	Recharge pulses followed by prolonged abstraction	Episodic contaminant flushing followed by concentration	Temporal oscillation between mobilization and stagnation	Planning uncertainty; resilience erosion	Dynamic allocation rules; conjunctive management

(continued)

Table 1.1 (continued)

Climate change driver	Key hydrological processes affected	Impacts on surface water systems	Impacts on groundwater systems	Water quality implications	Quality–quantity interaction mechanism	Implications for water sustainability	Adaptive management relevance
Increased climate variability and non-stationarity	Disruption of historical hydrological baselines	Unpredictable reservoir inflows and flow timing	Uncertain recharge and storage projections	Fluctuating pollutant concentrations	Breakdown of historical dilution assumptions	Higher planning uncertainty	Scenario-based modelling; adaptive governance
Land–atmosphere feedback intensification	Altered evapotranspiration and moisture recycling	Changes in regional rainfall recycling	Reduced long-term recharge stability	Gradual increase in dissolved solids	Long-term decline in flushing capacity	Chronic water quality deterioration	Integrated land–water planning
Surface water–groundwater decoupling	Reduced baseflow connectivity	Decreased environmental flows	Lower baseflow contribution; falling water tables	Reduced natural attenuation capacity	Loss of buffering between storage and flow systems	Reduced system resilience	Conjunctive use frameworks; recharge enhancement

pollutant attenuation while moderating runoff volumes (Kumar et al. 2024; Farhangi et al. 2025). However, implementation remains uneven across rapidly growing urban regions.

1.3.2 Anthropogenic Water Use and Its Implications

Industrial sectors exert dual pressures on water systems through high abstraction volumes and the release of pollutants. Water is widely used in manufacturing, cooling, mining, power generation, textile processing, chemical production, and the pharmaceutical industry. In many regions, abstraction rates strain local surface and groundwater reserves, particularly where regulatory oversight is weak (Lemessa et al. 2023). Industrial effluents may contain heavy metals, organic solvents, persistent organic pollutants, and synthetic compounds. Inadequate treatment facilitates the accumulation of contaminants in sediments and biota, posing long-term ecological and public health risks. Although cleaner production technologies, water recycling, and zero-liquid discharge systems are increasingly available, adoption varies significantly across sectors and regions (Radelyuk et al. 2023). Persistent contamination reduces the usable fraction of water resources, effectively constraining both water quality and available supply.

Agriculture remains the largest global freshwater user and a primary driver of groundwater extraction (Scanlon et al. 2007; Maghrebi et al. 2020). Expansion of irrigation infrastructure and widespread tube-well development have increased agricultural productivity, but frequently exceed natural recharge rates. Conventional irrigation methods, such as flood and furrow systems, exhibit low water-use efficiency, leading to evaporation losses, runoff, and deep percolation (Bwambale et al. 2022). In water-stressed regions, climate variability increases irrigation demand, accelerating aquifer depletion. Persistent overextraction lowers water tables, raises pumping energy requirements, and contributes to land subsidence and salinization, particularly in coastal and arid zones (Zaveri et al. 2016; Malki et al. 2017; Lu et al. 2025b). Although advanced irrigation technologies, such as drip and sprinkler systems, improve efficiency, economic and institutional barriers limit their widespread adoption (Koech and Langat 2018; Lakhiar et al. 2024).

1.3.3 Land-Use Change and Soil–Water Interactions

Conversion of forests, wetlands, and grasslands to agricultural or urban land alters infiltration, evapotranspiration, and runoff pathways. Reduced vegetation cover and soil compaction diminish water retention capacity, increase erosion, and limit groundwater recharge (Horel et al. 2022; Santana et al. 2023). Channel modification, river damming, and wetland drainage disrupt hydrological connectivity and reduce natural storage functions. Degradation of soil organic matter decreases infiltration efficiency and increases vulnerability to both drought and flooding. Conversely, conservation agriculture, agroforestry, and soil moisture management practices enhance infiltration and reduce erosion, thereby improving water-use efficiency and recharge potential (Pulighe et al. 2025; Sharma et al. 2025). Integrated land-use planning with water resource management is essential to mitigate hydrological imbalances.

Water quality degradation arises from point and non-point pollution sources associated with agriculture, industry, and domestic wastewater discharge. Nutrient enrichment from nitrogen and phosphorus loading promotes eutrophication, harmful algal blooms, oxygen depletion, and ecosystem restructuring when assimilative capacity is exceeded (Kronvang et al. 2020; Malone and Newton 2020). Climate variability can intensify these processes by elevating water temperatures and reducing dilution during low-flow conditions (Bonsdorff 2021).

Emerging contaminants, including pharmaceuticals, personal care products, endocrine-disrupting compounds, antibiotics, microplastics, and antimicrobial resistance genes, are increasingly detected in surface and groundwater systems. Many conventional wastewater treatment processes do not fully remove these compounds, allowing persistence and bioaccumulation (Yu et al. 2024; Cao et al. 2025). Extreme rainfall events enhance contaminant mobilization, while prolonged droughts concentrate dissolved constituents (Tong et al. 2022; Cao et al. 2025).

Groundwater contamination presents long-term sustainability challenges due to slow subsurface transport and limited natural attenuation. Agricultural nitrates, pesticides, industrial chemicals, landfill leachate, and saline intrusion degrade aquifer quality, often persisting for decades (Li et al. 2021; Qian et al. 2020; Gorelick and Zheng 2015). Restoration is technically complex and economically demanding, underscoring the importance of preventive management and monitoring (Burri et al. 2019).

1.4 Combined Effects of Climate and Human Activities

The interaction between climatic variability and anthropogenic interventions increasingly shapes water systems. Climate change alters the magnitude, timing, and variability of water supply, while human activities determine patterns of abstraction, storage, discharge, and pollutant loading. These pressures rarely operate independently; their interaction generates compounded, non-linear responses across hydrological and ecological systems (Haddeland et al. 2014). Hydro-climatic variability, particularly prolonged droughts and irregular precipitation, increases dependence on both surface water withdrawals and groundwater abstraction. During low-flow periods, reduced river discharge prompts intensified pumping from aquifers to meet agricultural, industrial, and municipal demands (Russo and Lall 2017). Where governance and monitoring are limited, abstraction frequently exceeds recharge, accelerating aquifer depletion. Declining groundwater levels reduce baseflow contributions to rivers, thereby reinforcing surface water scarcity and further concentrating pollutants under reduced dilution. This feedback loop weakens the traditional buffering role of groundwater storage (Soliman and Amin 2025). Under climate stress, systems that rely heavily on groundwater for drought resilience may experience long-term structural decline, reducing adaptive capacity.

Land-use change alters watershed hydrology and modifies system responses to climatic variability. Urban expansion and deforestation reduce infiltration and increase surface runoff, intensifying flood peaks during extreme rainfall events (Gashaw et al. 2018). Agricultural intensification reduces soil organic content and water retention, amplifying drought sensitivity. When combined with increased rainfall intensity and prolonged dry spells, these land-surface modifications exacerbate hydrological extremes. Reduced infiltration limits groundwater recharge during wet periods, while degraded soil moisture storage increases vulnerability during drought (Kayitesi et al. 2022). The loss of wetlands and floodplains further diminishes natural attenuation and storage capacity, heightening ecological and infrastructural risks.

Aquatic ecosystems depend on flow magnitude, timing, water temperature, and chemical composition. Climate-driven alterations in flow regimes interact with water abstraction, dam regulation, and pollution inputs to modify habitat structure and ecological connectivity (Mittal et al. 2016; Han et al. 2019; Huang et al. 2024). Reduced environmental flows, habitat fragmentation, and altered thermal regimes disrupt species migration, reproduction, and nutrient cycling (Brauer and Beheregaray 2020; Kuo et al. 2021). Elevated water temperatures combined with nutrient enrichment promote eutrophication and harmful algal blooms, reducing dissolved oxygen and altering community composition (Muruganandam et al. 2023; Woolway et al. 2022). Climate variability further intensifies these impacts by lowering dilution during drought and enhancing contaminant mobilization during floods, accelerating biodiversity loss and ecosystem homogenization (Kazmi et al. 2022).

The interaction of climatic and anthropogenic pressures affects ecosystem services that underpin human livelihoods. Declining water quality and altered flow regimes reduce fisheries productivity, impair natural water purification functions, and weaken flood regulation services (Albert et al. 2021). Communities dependent on aquatic resources are particularly vulnerable under combined hydrological and ecological stress. Floods and droughts disrupt infrastructure, agriculture, and energy production, generating economic losses and increasing public expenditure on disaster recovery (Dadson et al. 2017). In rural and coastal areas, degradation of wetlands and riverine systems reduces natural protection against storm surges and flooding, amplifying exposure to hazards (Hossain et al. 2018). These impacts are unevenly distributed, disproportionately affecting low-income and resource-dependent populations, thereby reinforcing social inequities (Darwall et al. 2018).

1.5 Implications for Water Sustainability

Sustainable water management requires maintaining hydrological reliability, water quality integrity, ecosystem functionality, and equitable access under conditions of climatic variability and increasing demand. The interaction of climate change and anthropogenic pressures has altered supply patterns, intensified contamination risks, and increased systemic uncertainty (Ciampittiello et al. 2024; Yin et al. 2025). These shifts expose structural weaknesses in allocation systems, institutional coordination, and risk preparedness.

1.5.1 Balancing Supply, Demand, and Ecological Thresholds

Climate-induced variability modifies the magnitude and seasonality of water availability, while demographic growth, agricultural expansion, and industrial development increase demand. This divergence complicates allocation decisions across domestic, agricultural, industrial, and environmental sectors. Reduced baseflows and declining groundwater levels constrain supply during peak demand periods, particularly in semiarid and densely populated basins. Sustaining environmental flows is essential to preserving riverine and wetland ecosystems that support nutrient cycling, habitat provision, and natural purification. However, ecological water requirements are frequently deprioritized during scarcity, accelerating ecosystem degradation and long-term productivity loss (Moghadam et al. 2022; Buri et al. 2025; Gunacti and Cetinkaya 2025). Over- and rigid-allocation regimes limit flexibility under climatic uncertainty, increasing exposure to hydrological extremes.

Institutional fragmentation remains a major barrier to sustainability. Responsibilities for water supply, irrigation, sanitation, environmental protection, and disaster management are often distributed across agencies with limited coordination (Sáez-Ardura et al. 2025). Inadequate regulatory enforcement, financial constraints, and limited technical capacity weaken the implementation of sustainable water policies (Hasan et al. 2023). Conventional planning frameworks frequently rely on historical hydrological data that do not reflect evolving climatic baselines. Without adaptive mechanisms that incorporate climate variability and emerging risks, management systems remain reactive rather than anticipatory (Akamani 2023; Dai et al. 2025). Strengthening institutional coherence and adaptive governance capacity is essential to sustain water systems under dynamic conditions.

1.5.2 Data Gaps, Uncertainty, and Socioeconomic Vulnerability

Effective water sustainability depends on reliable hydrological and water quality data. In many regions, monitoring networks are sparse, fragmented, or inconsistent, limiting the capacity to assess recharge rates, pollutant loads, and long-term trends (Sheffield et al. 2018; Koch et al. 2025). Climate change introduces additional uncertainty in precipitation intensity, extreme events, and land-use feedbacks, reducing confidence in traditional planning assumptions. Socioeconomic consequences of water system degradation extend beyond hydrology. Declining water availability and deteriorating quality increase public health risks, particularly in densely populated urban areas and informal settlements (Zimmermann et al. 2025; Miller et al. 2021). Agricultural productivity declines under water scarcity, threatening food security and rural livelihoods. Economic sectors dependent on a stable water supply, including agriculture, hydropower, industry, and tourism, face rising operational risks (Dadson et al. 2017; Zlati et al. 2024). Water insecurity disproportionately affects marginalized groups, including low-income households, women, and rural communities with limited adaptive capacity (Achore et al. 2020; Redhu and Jain 2024). Unequal exposure to scarcity, contamination, and climate-related hazards reinforces structural inequities and increases vulnerability to displacement and conflict. Table 1.2 presents key sustainability challenges associated with water system degradation, environmental, socioeconomic, and governance issues, highlighting their effects on sustaining ecosystems and socioeconomic activities.

1.6 Governance Transformation and Technological Innovation

Sustainable water systems depend on coordinated policy frameworks, institutional capacity, and technological advancement capable of responding to climatic non-stationarity and anthropogenic pressures. Technical interventions require enabling governance structures and regulatory coherence to be effective (Chowdhury and Behera 2022). Integrated Water Resources Management (IWRM) provides a framework for balancing economic development and ecological protection across sectors (Ngene et al. 2021), but implementation demands reduced institutional fragmentation among supply, irrigation, sanitation, environmental regulation, and disaster management authorities (Katusiime and Schütt 2020). Incorporating climate risk evaluation, environmental flow protection, and stakeholder participation strengthens responsiveness under evolving hydrological regimes (Akamani 2023).

Adaptive governance emphasizes flexibility, iterative learning, and reversible allocation strategies in the face of socioeconomic and climatic change (Dai et al. 2025). Effective adaptation requires regulatory enforcement, financial sustainability,

Table 1.2 Sustainability challenges and socioeconomic implications of water system degradation

Sustainability dimension	Key challenge	Primary drivers	Water quality–quantity interaction pathway	Systemic risk amplification mechanism	Socioeconomic impacts	Environmental consequences	Adaptive governance/ technical leverage points	Relevant SDGs	Key references
Water availability and security	Increasing scarcity and variability	Climate variability, over-abstraction, and land-use change	Reduced dilution capacity during low flows; declining recharge	Feedback between groundwater depletion and reduced baseflow	Increased water stress, allocation conflicts	Depleted rivers and aquifers	Conjunctive use, adaptive allocation rules, recharge enhancement	SDG 6, SDG 13	Garrick et al. (2020), Dadson et al. (2017)
Water quality degradation	Surface and groundwater contamination	Pollution, extreme events, reduced dilution	Flood-driven contaminant mobilization; drought concentration	Persistent contaminant accumulation in low-flow systems	Increased treatment costs, health burden	Eutrophication, hypoxia	Source control, advanced treatment, and monitoring networks	SDG 6, SDG 3	Malone and Newton (2020), Bonsdorff (2021), Burri et al. (2019)
Food and agricultural systems	Reduced productivity and irrigation stress	Water scarcity, inefficient irrigation, and groundwater mining	Salinity increases under declining recharge; nutrient runoff	Soil degradation reinforces hydrological stress	Yield loss, farmer income instability	Soil salinization, aquifer depletion	Efficient irrigation, managed aquifer recharge	SDG 2, SDG 6, SDG 12	Dolan et al. (2021), Gorelick and Zheng (2015)
Public health	Exposure to unsafe water	Contamination, sanitation failure, extreme events	Pathogen transport during floods; contaminant concentration in drought	Compounded disease risk under infrastructure failure	Waterborne diseases, healthcare costs	Contaminated water bodies	Safe sanitation, decentralized treatment	SDG 3, SDG 6	Brewis et al. (2020), Zimmermann et al. (2025)

(continued)

Table 1.2 (continued)

Sustainability dimension	Key challenge	Primary drivers	Water quality–quantity interaction pathway	Systemic risk amplification mechanism	Socioeconomic impacts	Environmental consequences	Adaptive governance/ technical leverage points	Relevant SDGs	Key references
Ecosystems and biodiversity	Loss of aquatic habitats	Altered flows, warming, and pollution	Thermal stress and nutrient enrichment under low flows	Biodiversity homogenization and trophic disruption	Decline in fisheries, cultural losses	Habitat loss, species decline	Environmental flow regulation, wetland restoration	SDG 14, SDG 15	Albert et al. (2021), Darwall et al. (2018)
Economic development	Infrastructure vulnerability	Floods, droughts, and resource degradation	Extreme-event induced infrastructure contamination	Rising recovery costs and productivity loss	Reduced economic growth	Degradation of natural capital	Climate-resilient infrastructure investment	SDG 8, SDG 9, SDG 13	Dadson et al. (2017), Ross et al. (2024)
Social equity	Unequal risk distribution	Governance gaps, poverty, marginalization	Reduced access to safe supply during scarcity	Disproportionate burden on vulnerable groups	Increased vulnerability and migration	Overexploitation of local resources	Inclusive governance, equitable allocation	SDG 5, SDG 10, SDG 16	Achore et al. (2020), Redhu and Jain (2024)
Governance and institutions	Weak adaptive capacity	Fragmented management, data gaps	Inadequate integration of quality and quantity planning	Policy inertia under non-stationarity	Inefficient allocation, delayed response	Poor ecosystem protection	IWRM, adaptive governance frameworks	SDG 6, SDG 16	Garrick et al. (2020), Mishra et al. (2021)
Energy–water nexus	Hydropower and cooling water vulnerability	Drought, temperature rise	Reduced flow availability and thermal limits	Competing sectoral allocation under scarcity	Energy supply instability	Altered thermal regimes	Integrated energy-water planning	SDG 7, SDG 13	Dadson et al. (2017)
Groundwater sustainability	Long-term aquifer degradation	Over-extraction, saline intrusion	Declining recharge and rising salinity	Irreversible storage loss	Higher pumping costs	Aquifer contamination and subsidence	Recharge protection, abstraction control	SDG 6, SDG 12	Gorelick and Zheng (2015), Burri et al. (2019)

transparency, and strengthened institutional capacity across transboundary and intersectoral systems (Hasan et al. 2023; Sáez-Ardura et al. 2025; Adom and Simatele 2022; Moreira et al. 2024; Whitley 2024). Technological innovation enhances efficiency, contaminant removal, and monitoring capacity. Advanced treatment technologies—including membrane filtration, oxidation processes, and biologically inspired systems—support reuse and quality management (Gupta et al. 2020). Smart water systems and digital tools such as remote sensing, Geographic Information Systems, sensor networks, and Artificial Intelligence improve operational control, real-time monitoring, and early-warning systems for floods and droughts (Gupta et al. 2020; Aivazidou et al. 2021). However, cost constraints and data governance challenges limit uniform adoption.

Improved prediction of hydrological extremes and contaminant transport remains essential (Hoffmann et al. 2020). Strengthening regional modelling that integrates climate drivers, land-use dynamics, and abstraction patterns enhances planning reliability (Hoffmann et al. 2020). Coupled surface water–groundwater modelling improves the assessment of recharge and aquifer sustainability (Conant et al. 2019), while expanded monitoring networks and the integration of remote sensing address data scarcity (Sheffield et al. 2018; Koch et al. 2025). Climate extremes modify contaminant transport pathways, with floods mobilizing stored pollutants and drought concentrating dissolved constituents (Johnston and Maher 2022; Bolan et al. 2024). Emerging contaminants, including microplastics and pharmaceuticals, present additional risks under altered hydrological conditions (Geris et al. 2022). Water sustainability challenges require integration of hydrological, ecological, socioeconomic, and institutional research domains (Zhao et al. 2021; Santos et al. 2023). Bridging the science–policy interface through knowledge co-production, decision-support tools, and comparative analysis strengthens adaptive planning and long-term resilience (Mishra et al. 2021). Figure 1.1 illustrates the integrated interaction between climate drivers, anthropogenic pressures, surface water–groundwater coupling, and central quality–quantity feedback mechanisms, highlighting cascading impacts and adaptive resilience pathways.

1.7 Conclusions

Water systems are experiencing profound transformation under the combined pressures of climate variability and anthropogenic intensification. Altered precipitation regimes, rising temperatures, cryospheric decline, and sea-level rise are modifying hydrological processes, while urban expansion, industrial activity, agricultural abstraction, land-use change, and pollutant discharge are reshaping storage dynamics and water quality conditions. These drivers interact across surface and groundwater systems, reducing natural buffering capacity, amplifying extremes, and accelerating ecological degradation. Hydrological non-stationarity, increased abstraction during drought, contaminant mobilization during floods, and declining environmental flows collectively undermine long-term reliability and resilience.

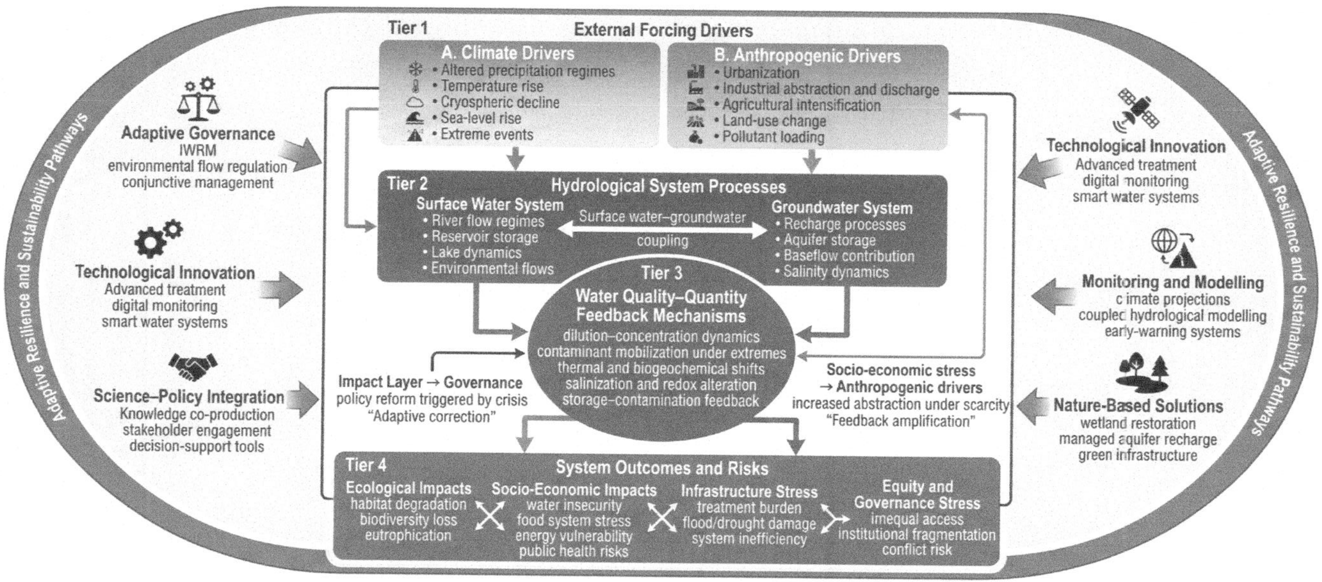

Fig. 1.1 Integrated systems framework illustrating the interaction between climate drivers and anthropogenic pressures, their influence on surface and groundwater processes, the central role of water quality–quantity feedback mechanisms, and the cascading ecological and socioeconomic impacts. The adaptive resilience envelope highlights governance, technological, modelling, and nature-based pathways for sustainable transformation of water systems

The resulting challenges extend beyond hydrology into ecosystem stability, economic productivity, public health, and social equity. Governance fragmentation, monitoring limitations, and uneven adaptive capacity further constrain effective responses under uncertainty. Sustaining water systems under accelerating climatic and socioeconomic change requires coordinated governance, strengthened data systems, technological advancement, and interdisciplinary integration capable of addressing both quantitative reliability and qualitative integrity. The systemic interactions outlined in this chapter provide the foundation for examining water quality–quantity linkages, predictive approaches, and resilience strategies in subsequent chapters.

References

Achore M, Bisung E, Kuusaana ED (2020) Coping with water insecurity at the household level: a synthesis of qualitative evidence. Int J Hyg Environ Health 230:113598. https://doi.org/10.1016/j.ijheh.2020.113598

Adom RK, Simatele MD (2022) The role of stakeholder engagement in sustainable water resource management in South Africa. Nat Resour Forum 46:410–427. https://doi.org/10.1111/1477-8947.12264

Agarwal V, Kumar M, Panday DP, Zang J, Munoz-Arriola F (2024) Unlocking the potential of remote sensing for arsenic contamination detection and management: challenges and perspectives. Curr Opin Environ Sci Health 100578. https://doi.org/10.1016/j.coesh.2024.100578

Agarwal V, Vishvendra Raj Singh B, Marsh S, Qin Z, Sen A, Kulhari K (2025) Integrated remote sensing for enhanced drought assessment: a multi-index approach in Rajasthan, India. Earth Space Sci 12(2):e2024EA003639. https://doi.org/10.1029/2024EA003639

Aivazidou E, Banias G, Lampridi M et al (2021) Smart technologies for sustainable water management: an urban analysis. Sustain 13. https://doi.org/10.3390/su132413940

Ajjur SB, Al-Ghamdi SG (2021) Evapotranspiration and water availability response to climate change in the Middle East and North Africa. Clim Change 166:1–18. https://doi.org/10.1007/s10584-021-03122-z

Akamani K (2023) The roles of adaptive water governance in enhancing the transition towards ecosystem-based adaptation. Water (Switzerland) 15. https://doi.org/10.3390/w15132341

Akhtar N, Syakir Ishak MI, Bhawani SA, Umar K (2021) Various natural and anthropogenic factors responsible for water quality degradation: a review. Water (Switzerland) 13. https://doi.org/10.3390/w13192660

Albert JS, Destouni G, Duke-Sylvester SM et al (2021) Scientists' warning to humanity on the freshwater biodiversity crisis. Ambio 50:85–94. https://doi.org/10.1007/s13280-020-01318-8

Allan RP, Barlow M, Byrne MP et al (2020) Advances in understanding large-scale responses of the water cycle to climate change. Ann N Y Acad Sci 1472:49–75. https://doi.org/10.1111/nyas.14337

Althoff D, Rodrigues LN, da Silva DD (2020) Impacts of climate change on the evaporation and availability of water in small reservoirs in the Brazilian Savannah. Clim Change 159:215–232. https://doi.org/10.1007/s10584-020-02656-y

Amanambu AC, Obarein OA, Mossa J et al (2020) Groundwater system and climate change: present status and future considerations. J Hydrol 589:125163. https://doi.org/10.1016/J.JHYDROL.2020.125163

Asif Z, Chen Z, Sadiq R, Zhu Y (2023) Climate change impacts on water resources and sustainable water management strategies in North America. Water Resour Manag 37:2771–2786. https://doi.org/10.1007/s11269-023-03474-4

Barbieri M, Barberio MD, Banzato F et al (2023) Climate change and its effect on groundwater quality. Environ Geochem Health 45:1133–1144. https://doi.org/10.1007/s10653-021-01140-5

Berghuijs WR, Collenteur RA, Jasechko S et al (2024) Groundwater recharge is sensitive to changing long-term aridity. Nat Clim Chang 144(14):357–363. https://doi.org/10.1038/s41558-024-01953-z

Bolan S, Padhye LP, Jasemizad T et al (2024) Impacts of climate change on the fate of contaminants through extreme weather events. Sci Total Environ 909. https://doi.org/10.1016/j.scitotenv.2023.168388

Bonsdorff E (2021) Eutrophication: early warning signals, ecosystem-level and societal responses, and ways forward: this article belongs to Ambio's 50th anniversary collection. theme: eutrophication. Ambio 50:753–758. https://doi.org/10.1007/s13280-020-01432-7

Brauer CJ, Beheregaray LB (2020) Recent and rapid anthropogenic habitat fragmentation increases extinction risk for freshwater biodiversity. Evol Appl 13:2857–2869. https://doi.org/10.1111/eva.13128

Brewis A, Workman C, Wutich A et al (2020) Household water insecurity is strongly associated with food insecurity: evidence from 27 sites in low- and middle-income countries. Am J Hum Biol 32:1–13. https://doi.org/10.1002/ajhb.23309

Buri ES, Keesara VR, Loukika KN et al (2025) Climate-adaptive optimal water resources management: a multi-sectoral approach for the Munneru river basin, India.. J Environ Manage 374:124014. https://doi.org/10.1016/j.jenvman.2024.124014

Burri NM, Weatherl R, Moeck C, Schirmer M (2019) A review of threats to groundwater quality in the anthropocene. Sci Total Environ 684:136–154. https://doi.org/10.1016/j.scitotenv.2019.05.236

Bwambale E, Abagale FK, Anornu GK (2022) Smart irrigation monitoring and control strategies for improving water use efficiency in precision agriculture: a review. Agric Water Manag 260:107324. https://doi.org/10.1016/j.agwat.2021.107324

Cao H, Liu J, You W et al (2025) Emerging contaminants in Taihu lake Basin: multi-dimensional prioritization, colloidal adsorption, and trophic-level disruptions. J Environ Manage 388:125965. https://doi.org/10.1016/j.jenvman.2025.125965

Chowdhury K, Behera B (2022) Institutional dynamics and water resource management: the case of traditional water bodies in West Bengal, India. Int J Water Resour Dev 38:836–860. https://doi.org/10.1080/07900627.2021.1916448

Ciampittiello M, Marchetto A, Boggero A (2024) Water resources management under climate change: a review. Sustain 16. https://doi.org/10.3390/su16093590

Conant B, Robinson CE, Hinton MJ, Russell HAJ (2019) A framework for conceptualizing groundwater-surface water interactions and identifying potential impacts on water quality, water quantity, and ecosystems. J Hydrol 574:609–627. https://doi.org/10.1016/j.jhydrol.2019.04.050

Dadson S, Hall JW, Garrick D et al (2017) Water security, risk, and economic growth: insights from a dynamical systems model. Water Resour Res 53:6425–6438. https://doi.org/10.1002/2017WR020640

Dai Y, Huang Z, Khan N, Labbo MS (2025) Smart water management: governance innovation, technological integration, and policy pathways toward economic and ecological sustainability. Water (Switzerland) 17:1–22. https://doi.org/10.3390/w17131932

Dao PU, Heuzard AG, Le TXH et al (2024) The impacts of climate change on groundwater quality: a review. Sci Total Environ 912:169241. https://doi.org/10.1016/j.scitotenv.2023.169241

Darwall W, Bremerich V, De Wever A et al (2018) The alliance for freshwater life: a global call to unite efforts for freshwater biodiversity science and conservation. Aquat Conserv Mar Freshw Ecosyst 28:1015–1022. https://doi.org/10.1002/aqc.2958

Deng C, Zhang H, Hamilton DP, Hollingsworth A (2025) Impacts of changes in climate and water demand on flow in a subtropical river catchment below a major dam. J Environ Manage 375:124137. https://doi.org/10.1016/J.JENVMAN.2025.124137

Dixit A, Madhav S, Mishra R, et al (2022) Impact of climate change on water resources, challenges and mitigation strategies to achieve sustainable development goals. Arab J Geosci 1514(15):1296. https://doi.org/10.1007/S12517-022-10590-9

Dolan F, Lamontagne J, Link R et al (2021) Evaluating the economic impact of water scarcity in a changing world. Nat Commun 121(12):1915-. https://doi.org/10.1038/s41467-021-22194-0

Eekhout JPC, Hunink JE, Terink W, De Vente J (2018) Why increased extreme precipitation under climate change negatively affects water security. Hydrol Earth Syst Sci 22:5935–5946. https://doi.org/10.5194/hess-22-5935-2018

Farhangi M, Hiteh M, Yousefyani H et al (2025) Consequences of urbanization on stormwater runoff and its management based on effective factors to choose the optimum scenario. J Environ Heal Sci Eng 23. https://doi.org/10.1007/s40201-025-00944-2

Garrick DE, Hanemann M, Hepburn C (2020) Rethinking the economics of water: an assessment. Oxford Rev Econ Policy 36:1–23. https://doi.org/10.1093/OXREP/GRZ035

Gashaw T, Tulu T, Argaw M, Worqlul AW (2018) Modeling the hydrological impacts of land use/land cover changes in the Andassa watershed, Blue Nile Basin, Ethiopia. Sci Total Environ 619–620:1394–1408. https://doi.org/10.1016/j.scitotenv.2017.11.191

Gavrilaș S, Burescu FL, Chereji BD, Munteanu FD (2025) The impact of anthropogenic activities on the catchment's water quality parameters. Water (Switzerland) 17:1–24. https://doi.org/10.3390/w17121791

Geris J, Comte JC, Franchi F et al (2022) Surface water-groundwater interactions and local land use control water quality impacts of extreme rainfall and flooding in a vulnerable semiarid region of Sub-Saharan Africa. J Hydrol 609:127834. https://doi.org/10.1016/j.jhydrol.2022.127834

Gorelick SM, Zheng C (2015) Global change and the groundwater management challenge. Water Resour Res 51:3031–3051. https://doi.org/10.1002/2014WR016825;PAGE:STRING:ARTICLE/CHAPTER

Goswami KB, Bisht PS (2017) The role of water resources in socio-economic development. Int J Res Appl Sci Eng Technol 5(XII):1669–1674

Gottlieb AR, Mankin JS (2024) Evidence of human influence on Northern Hemisphere snow loss. Nature 625:293–300. https://doi.org/10.1038/s41586-023-06794-y

Gunacti MC, Cetinkaya CP (2025) Integrated adaptive water allocation scenarios for wetland restoration: a case study of lake Marmara under climate change. Water (Switzerland) 17. https://doi.org/10.3390/w17131930

Gupta AD, Pandey P, Feijóo A et al (2020) Smart water technology for efficient water resource management: a review. Energies 13:1–23. https://doi.org/10.3390/en13236268

Haddeland I, Heinke J, Biemans H et al (2014) Global water resources affected by human interventions and climate change. Proc Natl Acad Sci U S A 111:3251–3256. https://doi.org/10.1073/pnas.1222475110

Häder DP, Banaszak AT, Villafañe VE et al (2020) Anthropogenic pollution of aquatic ecosystems: emerging problems with global implications. Sci Total Environ 713:136586. https://doi.org/10.1016/j.scitotenv.2020.136586

Han Z, Long D, Fang Y et al (2019) Impacts of climate change and human activities on the flow regime of the dammed Lancang River in Southwest China. J Hydrol 570:96–105. https://doi.org/10.1016/j.jhydrol.2018.12.048

Hasan N, Pushpalatha R, Manivasagam VS et al (2023) Global sustainable water management: a systematic qualitative review. Water Resour Manag 37:5255–5272. https://doi.org/10.1007/s11269-023-03604-y

Hoffmann S, Feldmann U, Bach PM et al (2020) A research agenda for the future of urban water management: exploring the potential of nongrid, small-grid, and hybrid solutions. Environ Sci Technol 54:5312–5322. https://doi.org/10.1021/acs.est.9b05222

Horel Á, Zsigmond T, Farkas C et al (2022) Climate change alters soil water dynamics under different land use types. Sustain 14. https://doi.org/10.3390/su14073908

Hossain MAR, Ahmed M, Ojea E, Fernandes JA (2018) Impacts and responses to environmental change in coastal livelihoods of south-west Bangladesh. Sci Total Environ 637–638:954–970. https://doi.org/10.1016/j.scitotenv.2018.04.328

Huang Y, Zhang K, Chao L et al (2024) Development of a comprehensive framework for quantifying the collective and individual influence of climate change and human activities on hydrological regimes. Ecol Indic 166:112487. https://doi.org/10.1016/j.ecolind.2024.112487

Jiménez-Bonilla A, Rodríguez-Rodríguez M, Yanes JL, Gázquez F (2025) Impact of climate change on permanent lakes in a semiarid region: Southwestern Mediterranean basin (S Spain). Sci Total Environ 961. https://doi.org/10.1016/j.scitotenv.2024.178305

Johnston SG, Maher DT (2022) Drought, megafires and flood—climate extreme impacts on catchment-scale river water quality on Australia's east coast. Water Res 218:118510. https://doi.org/10.1016/j.watres.2022.118510

Katusiime J, Schütt B (2020) Integrated water resources management approaches to improve water resources governance. Water 12:1–22. https://doi.org/10.3390/W12123424

Kayitesi NM, Guzha AC, Mariethoz G (2022) Impacts of land use land cover change and climate change on river hydro-morphology—a review of research studies in tropical regions. J Hydrol 615:128702. https://doi.org/10.1016/j.jhydrol.2022.128702

Kazmi SSUH, Wang YYL, Cai YE, Wang Z (2022) Temperature effects in single or combined with chemicals to the aquatic organisms: an overview of thermo-chemical stress. Ecol Indic 143:109354. https://doi.org/10.1016/j.ecolind.2022.109354

Koch H, Yangouliba GI, Liersch S (2025) From data scarcity to solutions: hydrological and water management modeling in a highly managed river basin. Water (Switzerland) 17. https://doi.org/10.3390/w17060823

Koech R, Langat P (2018) Improving irrigation water use efficiency: a review of advances, challenges and opportunities in the Australian context. Water (Switzerland) 10. https://doi.org/10.3390/w10121771

Kronvang B, Wendland F, Kovar K, Fraters D (2020) Land use and water quality. Water 12. https://doi.org/10.3390/W12092412

Kuang X, Liu J, Scanlon BR et al (2024) The changing nature of groundwater in the global water cycle. Science (80-)383:. https://doi.org/10.1126/science.adf0630

Kumar S, Kumar Vishwakarma R, Kumar Tyagi V et al (2024) Stormwater runoff characterization and adaptation of best management practices under urbanization and climate change scenarios. J Hydrol 635:131231. https://doi.org/10.1016/j.jhydrol.2024.131231

Kumar M, Panday DP, Bhagat C, Herbha N, Agarwal V, Jain V (2023) Demystifying the decadal shift in the extent of groundwater in the coastal aquifers of Gujarat, India: a case of reduced extent but increased magnitude of seawater intrusion. Sci Total Environ 165451. https://doi.org/10.1016/j.scitotenv.2023.165451

Kumar M, Saxena A, Tripathi S, Panday DP, Das N, Singh R, Mahlknecht J (2026) Localized or diffusive: insights into the source and geochemical dynamics of dissolved mercury contamination in groundwater of the mid-Gangetic Plains. npj Clean Water. https://doi.org/10.1038/s41545-026-00566-7

Kuo PH, Shih SS, Otte ML (2021) Restoration recommendations for mitigating habitat fragmentation of a river corridor. J Environ Manage 296:113197. https://doi.org/10.1016/j.jenvman.2021.113197

Lakhiar IA, Yan H, Zhang C et al (2024) A review of precision irrigation water-saving technology under changing climate for enhancing water use efficiency, crop yield, and environmental footprints. Agric 14. https://doi.org/10.3390/agriculture14071141

Lee M, Kim H, Lee JY, et al (2022) A shift towards integrated and adaptive water management in South Korea: building resilience against climate change. Water Resour Manag 365(36):1611–1625. https://doi.org/10.1007/S11269-022-03071-X

Lemessa F, Simane B, Seyoum A, Gebresenbet G (2023) Assessment of the impact of industrial wastewater on the water quality of rivers around the Bole Lemi Industrial Park (BLIP), Ethiopia. Sustain 15. https://doi.org/10.3390/su15054290

Leveque B, Burnet JB, Dorner S, Bichai F (2021) Impact of climate change on the vulnerability of drinking water intakes in a northern region. Sustain Cities Soc 66:102656. https://doi.org/10.1016/J.SCS.2020.102656

Li P, Karunanidhi D, Subramani T, Srinivasamoorthy K (2021) Sources and consequences of groundwater contamination. Arch Environ Contam Toxicol 80:1–10. https://doi.org/10.1007/s00244-020-00805-z

Lindner A, Stamm J, Lindner A, Stamm J (2025) Integrating climate change adaptation and water resource management: a critical overview. Stand 5(4). https://doi.org/10.3390/STANDARDS5010004

Lu J, Jia L, Menenti M et al (2025a) The impacts of drought on water availability: spatial and temporal analysis in the Belt and Road region (2001–2020). Int J Digit Earth 18. https://doi.org/10.1080/17538947.2025.2449706;WGROUP:STRING:PUBLICATION

Lu T, Luo P, Wang J et al (2025b) Soil salinity accumulation and groundwater degradation due to overexploitation over recent 40-year period in Yaoba Oasis. China. Soil Tillage Res 248:106398. https://doi.org/10.1016/j.still.2024.106398

Maghrebi M, Noori R, Bhattarai R et al (2020) Iran's agriculture in the Anthropocene. Earth's Futur 8. https://doi.org/10.1029/2020EF001547

Mahdian M, Hosseinzadeh M, Siadatmousavi SM et al (2023) Modelling impacts of climate change and anthropogenic activities on inflows and sediment loads of wetlands: case study of the Anzali wetland. Sci Rep 13:1–18. https://doi.org/10.1038/s41598-023-32343-8

Malki M, Bouchaou L, Hirich A et al (2017) Impact of agricultural practices on groundwater quality in intensive irrigated area of Chtouka-Massa, Morocco. Sci Total Environ 574:760–770. https://doi.org/10.1016/j.scitotenv.2016.09.145

Malone TC, Newton A (2020) The globalization of cultural eutrophication in the coastal ocean: causes and consequences. Front Mar Sci 7:1–30. https://doi.org/10.3389/fmars.2020.00670

McGrane SJ (2016) Impacts of urbanisation on hydrological and water quality dynamics, and urban water management: a review. Hydrol Sci J 61:2295–2311. https://doi.org/10.1080/02626667.2015.1128084

Miller JD, Workman CL, Panchang SV et al (2021) Water security and nutrition: current knowledge and research opportunities. Adv Nutr 12:2525–2539. https://doi.org/10.1093/advances/nmab075

Milner AM, Khamis K, Battin TJ et al (2017) Glacier shrinkage driving global changes in downstream systems. Proc Natl Acad Sci U S A 114:9770–9778. https://doi.org/10.1073/pnas.1619807114

Mishra BK, Kumar P, Saraswat C et al (2021) Water security in a changing environment: concept. Water 13:490

Mittal N, Bhave AG, Mishra A, Singh R (2016) Impact of human intervention and climate change on natural flow regime. Water Resour Manag 30:685–699. https://doi.org/10.1007/s11269-015-1185-6

Moghadam SH, Ashofteh PS, Loáiciga HA (2022) Optimal Water allocation of surface and ground water resources under climate change with WEAP and IWOA modeling. Water Resour Manag 36:3181–3205. https://doi.org/10.1007/s11269-022-03195-0

Moreira FD, Fonseca PRS, Miranda RM et al (2024) Stakeholder engagement for inclusive water governance in a rural community in Brazil. Front Water 6:1–10. https://doi.org/10.3389/frwa.2024.1378514

Mosley LM (2015) Drought impacts on the water quality of freshwater systems; review and integration. Earth-Sci Rev 140:203–214. https://doi.org/10.1016/j.earscirev.2014.11.010

Muruganandam M, Rajamanickam S, Sivarethinamohan S, Gaddam MKR, Velusamy P, Gomathi R, Muniasamy SK (2023) Impact of climate change and anthropogenic activities on aquatic ecosystem—a review. Environ Res 238:117233. https://doi.org/10.1016/J.ENVRES.2023.117233

Ngene BU, Nwafor CO, Bamigboye GO et al (2021) Assessment of water resources development and exploitation in Nigeria: a review of integrated water resources management approach. Heliyon 7:e05955. https://doi.org/10.1016/j.heliyon.2021.e05955

Nie Y, Pritchard HD, Liu Q, et al (2021) Glacial change and hydrological implications in the Himalaya and Karakoram. Nat Rev Earth Environ 2021 22 2:91–106. https://doi.org/10.1038/s43017-020-00124-w

Pahl-Wostl C (2020) Adaptive and sustainable water management: from improved conceptual foundations to transformative change. Int J Water Resour Dev 36:397–415. https://doi.org/10.1080/07900627.2020.1721268

Palmer PI, Wainwright CM, Dong B et al (2023) Drivers and impacts of Eastern African rainfall variability. Nat Rev Earth Environ 4:254–270. https://doi.org/10.1038/s43017-023-00397-x

Panday DP, Kumar M (2022) Climate indices and hydrological extremes: deciphering the best fit model. Environ Res 215:114301

Panday DP, Kumar M (2024) Decoding uranium variation over the Indian Peninsula through leveraging standardized precipitation evapotranspiration indices and groundwater level fluctuations. Groundw Sustain Dev 101241. https://doi.org/10.1016/J.GSD.2024.101241

Panday DP, Kumar M, Agarwal V, Torres-Martínez JA, Mahlknecht J (2024) Corroboration of arsenic variation over the Indian Peninsula through standardized precipitation evapotranspiration indices and groundwater level fluctuations: water quantity indicators for water quality prediction. Sci Total Environ 176339. https://doi.org/10.1016/j.scitotenv.2024.176339

Panday DP, Kumari A, Kumar M (2025) Alkalinity-salinity-sustainability: decadal groundwater trends and its impact on agricultural water quality in the Indian Peninsula. Sci Total Environ 978:179459. https://doi.org/10.1016/j.scitotenv.2025.179459

Pochodyła-Ducka E, Glińska-Lewczuk K, Jaszczak A (2023) Changes in stormwater quality and heavy metals content along the rainfall–runoff process in an urban catchment. Water 15. https://doi.org/10.3390/W15193505

Pulighe G, Belocchi A, Tani AC, Lupia F (2025) Impacts of agricultural management practices on water cycle, soil erosion and crop yields in a Mediterranean agroecosystem. Sci Total Environ 995:180111. https://doi.org/10.1016/j.scitotenv.2025.180111

Qian H, Chen J, Howard KWF (2020) Assessing groundwater pollution and potential remediation processes in a multi-layer aquifer system. Environ Pollut 263:114669. https://doi.org/10.1016/j.envpol.2020.114669

Radelyuk I, Klemeš JJ, Tussupova K (2023) Sustainable water use in industry—reasons, challenges, response of Kazakhstan. Circ Econ Sustain 3:2267–2283. https://doi.org/10.1007/s43615-023-00269-y

Redhu S, Jain P (2024) Unveiling the nexus between water scarcity and socioeconomic development in the water-scarce countries. Environ Dev Sustain 26:19557–19577. https://doi.org/10.1007/s10668-023-03425-4

Ross AG, Connolly K, Vögele S, Kuckshinrichs W (2024) A macro-level analysis of the socioeconomic impacts of climate change driven water scarcity: incorporating behavioural and resilience aspects. Water Res X 23:100223. https://doi.org/10.1016/J.WROA.2024.100223

Russo TA, Lall U (2017) Depletion and response of deep groundwater to climate-induced pumping variability. Nat Geosci 10:105–108. https://doi.org/10.1038/ngeo2883

Sadok W, Lopez JR, Smith KP (2021) Transpiration increases under high-temperature stress: potential mechanisms, trade-offs and prospects for crop resilience in a warming world. Plant Cell Environ 44:2102–2116. https://doi.org/10.1111/pce.13970

Sáez-Ardura F, Parra-Salazar M, Vallejos-Romero A et al (2025) Exploring the socio-environmental regulation of water—a systematic review of sustainable watershed management. Sustain 17:1–24. https://doi.org/10.3390/su17041588

Santana MLT, de Faria VL, Barbosa SM et al (2023) Impacts of land use changes on soil functions and water security: insights from a three-year-long study in the cantareira system. Southeast of Brazil. Sustain 15. https://doi.org/10.3390/su151813395

Santos E, Carvalho M, Martins S (2023) Sustainable water management: understanding the socioeconomic and cultural dimensions. Sustain 15:1–21. https://doi.org/10.3390/su151713074

Scanlon BR, Jolly I, Sophocleous M, Zhang L (2007) Global impacts of conversions from natural to agricultural ecosystems on water resources: quantity versus quality. Water Resour Res 43. https://doi.org/10.1029/2006WR005486

Sharma R, Rallapalli S, Magner J (2025) Optimizing water-efficient agriculture: evaluating the sustainability of soil management and irrigation synergies using fuzzy extent analysis. Sci Rep 15:1–23. https://doi.org/10.1038/s41598-025-15426-6

Sheffield J, Wood EF, Pan M et al (2018) Satellite remote sensing for water resources management: potential for supporting sustainable development in data-poor regions. Water Resour Res 54:9724–9758. https://doi.org/10.1029/2017WR022437

Soliman K, Amin D (2025) Vulnerability of groundwater quantity for an arid coastal aquifer under the climate change and extensive exploitation. Appl Water Sci 15:1–19. https://doi.org/10.1007/s13201-025-02404-6

Soranno PA, Hanly PJ, Webster KE et al (2025) Abrupt changes in algal biomass of thousands of US lakes are related to climate and are more likely in low-disturbance watersheds. Proc Natl Acad Sci U S A 122:e2416172122. https://doi.org/10.1073/PNAS.2416172122;PAGE:STRING:ARTICLE/CHAPTER

Tabari H (2020) Climate change impact on flood and extreme precipitation increases with water availability. Sci Rep 10:1–10. https://doi.org/10.1038/s41598-020-70816-2

Tariq ZH, Willis HH, Tariq ZH, Willis HH (2024) Examining trends in the food–energy–water security nexus and its relationships with human development, population growth, and conflict. Sustain 16:16:. https://doi.org/10.3390/SU16188255

Thomas J, Joseph S (2024) Impacts of recent rainfall changes on agricultural productivity and water resources within the Southern Western Ghats of Kerala, India. Environ Monit Assess 196:. https://doi.org/10.1007/s10661-023-12270-x

Tong X, Mohapatra S, Zhang J et al (2022) Source, fate, transport and modelling of selected emerging contaminants in the aquatic environment: current status and future perspectives. Water Res 217:118418. https://doi.org/10.1016/j.watres.2022.118418

Wang X, Liu L (2023) The impacts of climate change on the hydrological cycle and water resource management. Water (Switzerland) 15:2–5. https://doi.org/10.3390/w15132342

Whitley H (2024) Exogenous, endogenous, and peripheral actors: a situational analysis of stakeholder inclusion within transboundary water governance. Sustain 16. https://doi.org/10.3390/su16093647

Woolway RI, Sharma S, Smol JP (2022) Lakes in hot water: the impacts of a changing climate on aquatic ecosystems. Bioscience 72:1050–1061. https://doi.org/10.1093/biosci/biac052

Wu F, Yang X, Cui Z et al (2024a) The impact of human activities on blue-green water resources and quantification of water resource scarcity in the Yangtze River Basin. Sci Total Environ 909:168550. https://doi.org/10.1016/j.scitotenv.2023.168550

Wu Y, Yin X, Zhou G et al (2024b) Rising rainfall intensity induces spatially divergent hydrological changes within a large river basin. Nat Commun 15. https://doi.org/10.1038/s41467-023-44562-8

Yin W, Yang X, Liu W (2025) Sustainable management and regulation of agricultural water resources in the context of global climate change. Sustain 17:1–5. https://doi.org/10.3390/su17062760

Yu Y, Wang Z, Yao B, Zhou Y (2024) Occurrence, bioaccumulation, fate, and risk assessment of emerging pollutants in aquatic environments: a review. Sci Total Environ 923:171388. https://doi.org/10.1016/j.scitotenv.2024.171388

Zaveri E, Grogan DS, Fisher-Vanden K et al (2016) Invisible water, visible impact: groundwater use and Indian agriculture under climate change. Environ Res Lett 11. https://doi.org/10.1088/1748-9326/11/8/084005

Zhao D, Liu J, Sun L et al (2021) Quantifying economic-social-environmental trade-offs and synergies of water-supply constraints: an application to the capital region of China. Water Res 195:116986. https://doi.org/10.1016/j.watres.2021.116986

Zimmermann K, Abadi AM, Brauman KA et al (2025) Addressing water scarcity to support climate resilience and human health. Integr Environ Assess Manag 21:291–300. https://doi.org/10.1093/inteam/vjaf001

Zlati ML, Antohi VM, Ionescu RV et al (2024) Quantyfing the impact of the water security index on socioeconomic development in EU27. Socioecon Plann Sci 93:101912. https://doi.org/10.1016/J.SEPS.2024.101912

Zou XY, Peng XY, Zhao XX, Chang CP (2023) The impact of extreme weather events on water quality: international evidence. Nat Hazards 115:1–21. https://doi.org/10.1007/s11069-022-05548-9

Chapter 2
Interlinkages Amongst Water Quality and Quantity, Humanity, and Sustainability

Abstract Water security in the twenty-first century is increasingly threatened by declining water availability and worsening water quality driven by climatic variability, land-use change, and population growth. Understanding coupled water quantity-quality (QnQ) dynamics, is therefore essential for resilient management. This review traces the evolution of hydrological and hydrochemical models—from early foundations such as Streeter–Phelps and rainfall–runoff formulations to widely used tools including SWAT, MIKE SHE, INCA, and QUAL2E—highlighting how conceptual, semi-distributed, and integrated frameworks have improved representation of flow, contaminant transport, and biogeochemical processes. Despite progress, key barriers remain, particularly data scarcity, parameter uncertainty, and limited treatment of extremes, which restrict model transferability across scales. The review emphasizes that incorporating climatic indices, especially the Standardized Precipitation Index (SPI), strengthens QnQ coupling by linking hydroclimatic anomalies with contaminant dynamics. A comparison of global nitrate concentrations with SPI-9 shows that drought conditions are associated with elevated nitrate accumulation, whereas wetter periods promote dilution and aquifer recharge, underscoring climate sensitivity in groundwater quality. Beyond process-based modelling, the review highlights the growing impact of machine-learning-assisted approaches (e.g., Random Forest, XGBoost, Support Vector Machines, and Long Short-Term Memory networks) for capturing nonlinear, multi-scale relationships among hydrological, climatic, and hydrochemical variables. When integrated with human and sustainability dimensions, these tools move from prediction to decision support for equitable and adaptive governance. The proposed Integrated Machine Learning–Humanity–Sustainability Framework positions QnQ analysis within a socio-hydrological and sustainability context, showing that combining climatic indices, hydrochemical data, and modern computational tools offers a robust pathway for managing interconnected water systems under climatic and anthropogenic stress.

Keywords Water quantity–quality interaction · Hydroclimatic variability · Machine learning · Integrated modelling · Sustainability framework · Coupled human–water systems

M. Kumar et al., *Bridging Water Quality and Quantity in the Anthropocene*, SpringerBriefs in Water Science and Technology,
https://doi.org/10.1007/978-3-032-25655-3_2

List of Abbreviations

ABM	Agent-Based Model
AMS	American Meteorological Society
ASABE	American Society of Agricultural and Biological Engineers
BOD	Biochemical Oxygen Demand
CEH	Centre for Ecology & Hydrology (United Kingdom)
DHI	Danish Hydraulic Institute
DO	Dissolved Oxygen
EPA	Environmental Protection Agency (United States)
FAO	Food and Agriculture Organization of the United Nations
GWP	Global Water Partnership
HEC	Hydrologic Engineering Center (U.S. Army Corps of Engineers)
HBV	Hydrologiska Byråns Vattenbalansavdelning (Hydrological Bureau Water Balance Model, Sweden)
HSPF	Hydrological Simulation Program—FORTRAN
HYPE	Hydrological Predictions for the Environment
IHP	International Hydrological Programme (UNESCO)
INCA	Integrated Catchment Model
IWRM	Integrated Water Resources Management
IDB	Inter-American Development Bank
JRC	Joint Research Centre (European Commission)
MIKE SHE / MIKE 11	Integrated Hydrological Modelling System developed by DHI
MODFLOW	Modular Finite-Difference Groundwater Flow Model (USGS)
MT3DMS	Modular Three-Dimensional Multispecies Transport Model
NGO	Non-Governmental Organization
OECD	Organisation for Economic Co-operation and Development
OpenMI	Open Modelling Interface (European Modelling Standard)
PDSI	Palmer Drought Severity Index
PCR-GLOBWB	PCRaster Global Water Balance Model
QnQ	Quantity and Quality (in Water Studies)
SDG	Sustainable Development Goal
SDGs	Sustainable Development Goals (UN framework)
SEAWAT	Variable-Density Groundwater Flow and Transport Model
SEI	Stockholm Environment Institute
SHETRAN	Distributed River Basin Flow and Transport Modelling System
SMHI	Swedish Meteorological and Hydrological Institute

SPI	Standardized Precipitation Index
SPEI	Standardized Precipitation Evapotranspiration Index
SWAT	Soil and Water Assessment Tool
SWAT+	An extended version of the SWAT, including human and policy modules
SWMM	Storm Water Management Model
UNESCO	United Nations Educational, Scientific, and Cultural Organization
UNICEF	United Nations International Children's Emergency Fund
UN	United Nations
USACE	United States Army Corps of Engineers
USEPA	United States Environmental Protection Agency
USGS	United States Geological Survey
WEP	Water–Energy–Food–Climate Policy Model (UNESCO–IHP)
WEAP	Water Evaluation and Planning System
WHO	World Health Organization
WQ	Water Quality
WQI	Water Quality Index
WWF	World Wide Fund for Nature

2.1 Introduction

Water systems worldwide face coupled stresses in availability and quality from climate variability and intensifying human pressures. Low flows concentrate contaminants and reduce assimilative capacity, while high-intensity rainfall accelerates runoff, erosion, and non-point nutrient and sediment delivery—showing quantity and quality cannot be treated separately. Addressing this requires looking to the evolution of modelling, the climate signals structuring hydrologic and biogeochemical variability, and the human–sustainability dimensions that determine who benefits, who bears risk, and how ecosystems fare (Kumar et al. 2026a; Kumar et al. 2026b; Beven 2012; Vörösmarty et al. 2010).

Modelling has advanced from early conceptualizations to coupled frameworks, yet persistent blind spots remain. For quantity, parsimonious conceptual models such as HBV enabled useful discharge simulation in data-limited settings, while semi-distributed SWAT improved representation of land management, soil–plant processes, and basin heterogeneity, and integrated surface–subsurface platforms such as MIKE SHE increased process fidelity at higher data and computational cost (Lindström et al. 1997; Arnold et al. 2012; Panday et al. 2025). For quality, the Streeter–Phelps oxygen-sag model and 1-D river tools such as QUAL2E formalized mass balance and reaction kinetics, later extended by semi-distributed frameworks (e.g., INCA) and hydrodynamic–water-quality systems that better resolve spatio-temporal variability (Streeter and Phelps 1925; Brown and Barnwell 1987; Whitehead et al. 2009; Kumar et al. 2025a). Composite indices like WQI helped communicate status but raised concerns about weighting and loss of diagnostic detail (Sutadian et al. 2016). Across both domains, structural and parameter uncertainty, scale mismatches, limited representation of compound extremes, and weak coupling between flow regimes and pollutant dynamics remain recurring challenges (Beven 2012; Blöschl et al. 2013; Kumar et al. 2025b).

Climate signals can organize variability in both quantity and quality, but integration into coupled models is uneven. The Standardized Precipitation Index (SPI) summarizes precipitation anomalies across timescales and supports the detection and screening of drought and wet spells (McKee et al. 1993). Because SPI is precipitation-only, it may understate stress under warming-driven evaporative demand, motivating complementary metrics such as SPEI and, critically, the explicit propagation of these signals through hydrologic pathways that affect dilution, residence time, reaction rates, and non-point pollutant mobilization (Vicente-Serrano et al. 2010). Climate-aware coupling explains why identical loads yield different concentrations under contrasting flow states and why objectives must be stress-tested against droughts and pluvial extremes rather than means (Tripathi et al. 2025; Vörösmarty et al. 2010; Farooq et al. 2024; Panday et al. 2024; Panday and Kumar 2024).

Human and sustainability dimensions complete the picture. Urban, industrial, and irrigation withdrawals reshape hydrographs and thermal regimes; land-use change alters nutrient and sediment sources and pathways; and governance allocates risks and benefits, often affecting equity. Achieving safe, sufficient water while maintaining ecosystem integrity requires treating human drivers as endogenous and linking outputs to policy levers, distributional outcomes, and ecological thresholds aligned with SDGs (Vörösmarty et al. 2010; UNESCO 2020). Yet many studies still separate quantity and quality, use climate indices mainly for monitoring rather than predictive coupling, rely on sparse concentration data, and propagate uncertainty inconsistently from forcings and parameters to decisions (McKee et al. 1993; Vicente-Serrano et al. 2010; Blöschl et al. 2013). These gaps motivate a reframed inquiry into modelling evolution, climate controls on Q–Q linkages, and bringing humanity and sustainability from the margins to the core (Lindström et al. 1997; Arnold et al. 2012; Beven 2012; Whitehead et al. 2009).

Guided by this rationale, the chapter pursues three interlocking objectives. It synthesizes the evolution of quantity and quality modelling—conceptual, semi-distributed, and fully coupled developments—along with data demands, uncertainty treatment, and decision relevance. It explains how climate signals, focusing on SPI and related indices, structure water quality–quantity linkages by tracing the propagation of drought and wet spells through hydrological and biogeochemical processes that govern dilution, transport, and transformation. Finally, it interrogates humanity and sustainability by representing water use and land-system change as dynamic drivers and translating outputs into implications for equity, health, ecosystem integrity, and resilient integrated water-resources management.

2.2 Trends and Evolution in Water Quantity and Quality Modelling

2.2.1 *Evolution of Water Quality Models*

The earliest models of water quality were rooted in relatively simple mass-balance concepts. The classic Streeter–Phelps formulation described the biochemical oxygen demand (BOD) and dissolved oxygen (DO) sag curves in polluted rivers and was among the first attempts to represent the fate of pollution in aquatic systems mathematically (Streeter and Phelps 1925). This framework was extended through models such as QUAL2E, which expanded the number of constituents represented—nutrients, algal growth, temperature, and oxidation–reduction reactions—making it highly influential in regulatory contexts (Brown and Barnwell 1987). Over time, the limitations of these one-dimensional river models led to the development of semi-distributed catchment frameworks such as the Integrated Catchment Model (INCA), which explicitly linked terrestrial nutrient sources with aquatic processes and allowed representation of both point and diffuse pollution (Whitehead et al. 2009). More recently, hydrodynamic–water quality tools such as MIKE 11 and HydroPol2D have offered improved spatial resolution and the ability to simulate pollutant transport under stormwater dynamics and variable flow regimes (Wang et al. 2020). Alongside these process-based models, composite indices such as the Water Quality Index (WQI) emerged to simplify complex chemical datasets into single values for communication and monitoring (Sutadian et al. 2016). While effective for public reporting, WQI sacrifices diagnostic depth and often masks underlying processes. Despite this evolution, water quality models continue to face recurring challenges: sparse concentration datasets for calibration, poor representation of non-point sources, and difficulties scaling up to entire catchments (Whitehead et al. 2009; Beven 2012).

2.2.2 *Evolution of Water Quantity Models*

In parallel, water quantity modelling has undergone its own trajectory of conceptual broadening and technical sophistication. Conceptual models such as the HBV family offered parsimonious approaches, relying on precipitation and temperature as primary inputs to simulate runoff, evapotranspiration, and snowmelt processes in lumped or semi-distributed formats (Lindström et al. 1997). These models became widely used in operational hydrology due to their low data requirements and robustness under constrained conditions. The subsequent generation of semi-distributed models, exemplified by the Soil and Water Assessment Tool (SWAT), added considerable functionality by incorporating land use, soil characteristics, agricultural practices, and water management interventions at the sub-basin scale (Arnold et al. 2012). Fully integrated, physically based frameworks such as MIKE SHE and SHETRAN marked another step forward, representing surface–subsurface interactions and groundwater dynamics at fine resolutions (Refsgaard and Storm 1995).

On global and continental scales, models like WaterGAP and PCR-GLOBWB have further expanded the field by integrating human water use, reservoir operations, and large-scale climate variability into hydrological simulations. Despite this remarkable progress, water quantity models still struggle with parameter and structural uncertainty, inconsistent representation of extreme events, and mismatches in scale between model grids and management units (Beven 2012; Blöschl et al. 2013). These limitations constrain their direct applicability to integrated water-resource management and sustainability assessments.

2.2.3 Integration of Climate Indices into Water Quality–Quantity Studies

The integration of climatic indicators with hydrochemical datasets provides a vital perspective for understanding the coupled dynamics of **water quantity and quality (QnQ)** at multiple spatial and temporal scales. Numerous studies have highlighted that the variability in precipitation, temperature, and evapotranspiration strongly influences the chemical evolution of groundwater systems (Taylor et al. 2013; Scanlon et al. 2012). Among the various hydroclimatic indices, the **Standardized Precipitation Index (SPI)** serves as a robust measure for quantifying meteorological droughts and wetness periods (McKee et al. 1993), and its correlation with water quality variables, such as nitrate, salinity, and organic carbon, reveals critical insights into the **climate–contaminant nexus** (Wang et al. 2020; Kaushal et al. 2018).

Understanding this interplay is essential for predicting how aquifer quality responds to prolonged climate anomalies, recharge fluctuations, and anthropogenic stressors, all of which govern the sustainability of groundwater resources globally (Bonsor et al. 2018). Figure 2.1 illustrates this integration using the global distribution of nitrate (NO_3-) concentrations in groundwater (panel a) and the nine-month Standardized Precipitation Index (SPI-9) representing hydroclimatic variability (panel b). Elevated nitrate concentrations occur in regions characterized by intensive agricultural activity, groundwater over-extraction, and fertilizer leaching—particularly across parts of India, China, Europe, and North America (Zhang et al. 2018; Oelsner and Stets 2019). Conversely, regions showing moderate to high SPI-9 values—indicative of wet periods—tend to exhibit lower nitrate levels, reflecting enhanced dilution and aquifer recharge.

The climatic control on nitrate dynamics becomes even more apparent when considering time lags and feedbacks. During prolonged drought episodes (negative SPI), evapo-concentration increases ion strength and redox potential, promoting nitrate persistence and accumulation in the unsaturated zone (Scanlon et al. 2007). Subsequent rainfall events following dry periods can trigger nitrate flushing and temporary spikes in groundwater contamination (Wang et al. 2020). Such episodic

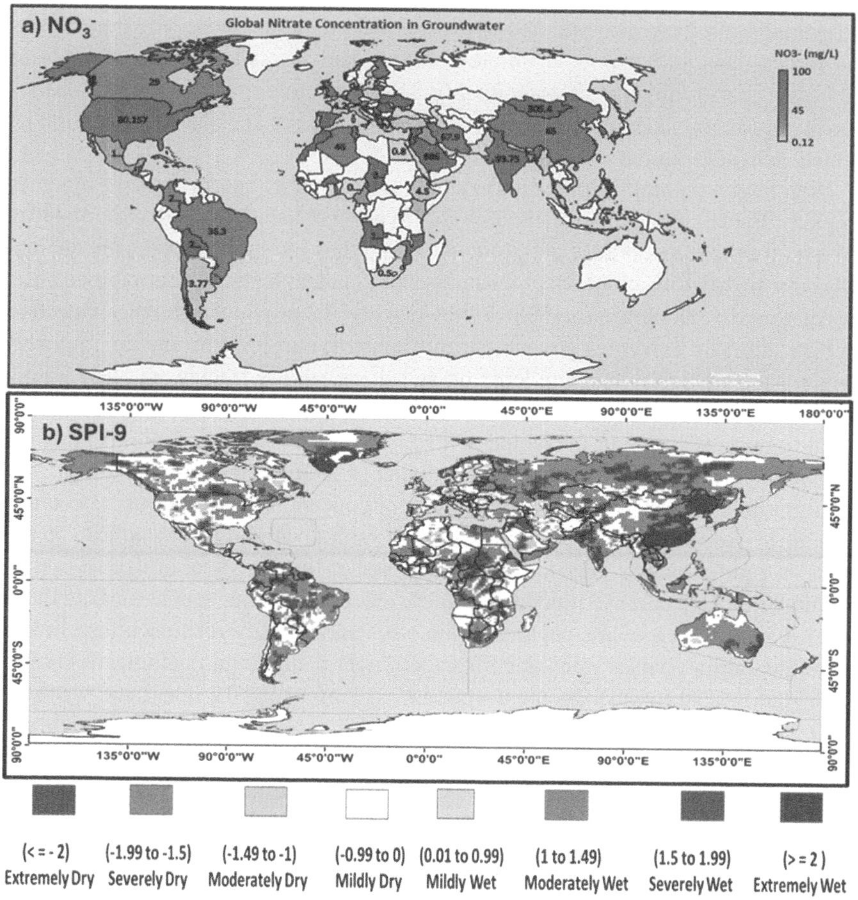

Fig. 2.1 Global Interaction between Water Quantity and Quality Indicators: **a** Nitrate Concentration (NO_3-) in groundwater and **b** Nine-month Standardized Precipitation Index (SPI-9). The maps illustrate the spatial coupling between hydrochemical contamination and climatic variability. Regions experiencing prolonged droughts (negative SPI) exhibit higher nitrate accumulation, while wetter zones (positive SPI) show enhanced dilution

yet recurrent behavior underscores the **nonlinear response** of aquifer systems to climate forcing, reinforcing the need for integrative analytical frameworks that couple hydrology and hydrochemistry under dynamic climatic conditions (Taylor et al. 2013).

Beyond empirical observation, coupling SPI and water-quality data with machine learning (ML) models enables deeper, data-driven insights. Algorithms such as **Long Short-Term Memory (LSTM)** networks can model temporal dependencies between rainfall deficits and delayed nitrate responses. At the same time, **Random Forest (RF)** and **Extreme Gradient Boosting (XGBoost)** can evaluate the relative influence of climatic, land-use, and anthropogenic variables on contaminant occurrence. By integrating both hydroclimatic and hydrochemical datasets, these ML frameworks can forecast spatial hotspots of nitrate pollution and assess how future drought or wetness scenarios might alter water quality. This predictive coupling enhances early-warning systems, enabling proactive water management and the identification of climate-sensitive aquifers.

Therefore, integrating climate indices with water-quality data bridges the methodological divide among hydroclimatology, hydrogeochemistry, and sustainability science. It transforms traditional groundwater studies into holistic assessments that reflect the real-world complexity of water systems under change. The combined analysis of nitrate concentration and SPI-9 shown in Fig. 2.1 provides a global perspective on **how climatic extremes govern contaminant dynamics**, forming an empirical basis for the advanced machine learning and sustainability frameworks that follow in subsequent sections.

The empirical synthesis presented in Fig. 2.1 thus provides a pivotal link between observed hydroclimatic variability and evolving groundwater quality conditions. By aligning climatic indices such as SPI with nitrate concentration dynamics, it becomes possible to trace the cumulative imprint of both natural variability and anthropogenic stress on aquifer health. This connection establishes the analytical bridge from data interpretation to predictive modelling, where machine learning can assimilate these coupled signals to forecast contamination risks and identify resilience thresholds. Such integration enables translating hydroclimatic patterns into actionable knowledge, guiding adaptive management strategies and informing the Integrated Machine Learning–Humanity–Sustainability Framework discussed in the subsequent sections. Table 2.1 synthesizes the progression from classic, one-way river water-quality models (e.g., Streeter–Phelps; QUAL/QUAL2E) and hydrology-only rainfall–runoff tools (e.g., HBV, TOPMODEL) to fully coupled watershed, groundwater, and integrated socio-hydrological platforms (e.g., SWAT, INCA, MIKE SHE/MIKE 11, MODFLOW–MT3DMS, WEAP, WaterGAP, SWAT+). It emphasizes how modelling capability expands with spatial resolution and feedback complexity—from flow-driven quality simulations to two-way solute–flow interactions and human–environment linkages—while pointing to established repositories and validated basin applications for reproducible deployment.

2.3 Humanity and Sustainability Dimensions of Water Quality–Quantity Linkages

The interdependence of water quality and quantity is not only a scientific problem but a profoundly social and sustainability challenge. Hydrological changes are ultimately mediated through human use, management decisions, and institutional frameworks, while water quality outcomes directly affect health, equity, and ecological integrity. To fully capture these dimensions, modelling and assessment efforts must extend

Table 2.1 Comparative overview of widely used water-quantity, water-quality, and coupled QnQ models, summarizing their origins, process representation, scale, and integration level. The table also highlights common data repositories, typical applications, and key validation references to guide model selection

Model and origin	Primary focus/process representation	Model category	Scale/spatial resolution	Integration of quantity and quality	Data source/repository	Typical applications and basin examples	Validated reference
Streeter–Phelps Model (U.S. Public Health Service 1925)	Dissolved oxygen balance and biochemical oxygen demand decay in rivers	Water quality	Reach-scale	Purely hydrodynamic input (no feedback)	U.S. EPA Model Archive	River DO–BOD simulation (e.g., Ohio River)	Streeter and Phelps (1925)
QUAL/ QUAL2E/ QUAL2Kw (USEPA 1970s–2000s)	Multi-constituent river quality: nutrients, DO, and temperature dynamics	Water quality	River network (1-D)	One-way coupling from flow to quality	USEPA BASINS Platform	River regulation and WQ standards in the U.S. and Asia	Brown and Barnwell (1987)
WASP/ CE-QUAL-W2 (USEPA and USACE 1980s)	Eutrophication, toxics, sediment–water interactions, temperature	Water quality	1-D/2-D reservoir and river systems	Hydrodynamic–biogeochemical coupling	USACE HEC/ EPA WASP repositories	Lakes Erie and Ontario, Chesapeake Bay	Cole and Wells (2006)
HBV (SMHI, Sweden 1972)	Rainfall–runoff processes, snowmelt dynamics	Water quantity	Catchment (semi-distributed)	None (hydrology only)	SMHI Hydrology Database	Nordic and Alpine basins	Lindström et al. (1997)

(continued)

Table 2.1 (continued)

Model and origin	Primary focus/process representation	Model category	Scale/spatial resolution	Integration of quantity and quality	Data source/ repository	Typical applications and basin examples	Validated reference
SWMM (USEPA 1971)	Urban stormwater runoff, sewer hydraulics, pollutant wash-off	Quantity + quality	Urban catchment	Event-based flow–pollutant coupling	USEPA SWMM Repository	Urban drainage planning globally	Rossman (2010)
HSPF (USGS—EPA 1980)	Continuous watershed hydrology and water quality	Coupled	Sub-basin distributed	Fully coupled rainfall–runoff–pollutant transport	USEPA BASINS System	Chesapeake Bay Program	Bicknell et al. (1985)
TOPMODEL (Beven and Kirkby 1979)	Variable source area runoff, topographic control of flow	Hydrological	Catchment	None	Lancaster University Repository	U.K., European basins	Beven and Kirkby (1979), Beven (2012)
SWAT (USDA ARS 1995)	Land-use, hydrology, sediment, and nutrient cycling	Coupled	Sub-basin distributed	Fully integrated hydrology–WQ–land use	SWAT Database (TAMU)	Mississippi, Ganges, Danube	Arnold et al. (2012)
INCA (UK CEH 1998)	Nitrogen, phosphorus, carbon, and dissolved oxygen cycles	Coupled	Reach-scale semi-distributed	Flow and nutrient processes are dynamically linked	UK CEH INCA Portal	Thames, Elbe, Danube	Whitehead et al. (2009)
MIKE SHE/ MIKE 11 (DHI 1990s)	Fully distributed surface–subsurface hydrology with water quality	Coupled	Grid-based catchment	Two-way dynamic coupling between flow and solute transport	DHI Repository	Rhine, Mekong	Graham and Butts (2005)

(continued)

Table 2.1 (continued)

Model and origin	Primary focus/process representation	Model category	Scale/spatial resolution	Integration of quantity and quality	Data source/ repository	Typical applications and basin examples	Validated reference
MODFLOW + MT3DMS (USGS 1990s)	Groundwater flow and solute transport	Coupled	3-D aquifer	Two-way groundwater–solute feedback	USGS MODFLOW Portal	Ogallala, Nile Deltas	Zheng and Wang (1999)
WaterGAP (Frankfurt Univ. 2003)	Global water balance and water use	Hydrological + Socio-economic	0.5° global grid	Simplified flow–use feedback	JRC / WWF WaterGAP DB	Global water stress mapping	Döll et al. (2003)
WEAP (SEI 2005)	Policy-driven integrated water resource planning	Socio-hydrological	Basin/national	Scenarios link quality modules externally	SEI WEAP DB	Jordan, Zambezi, Nile	Yates et al. (2005)
WEP (UNESCO-IHP 2006)	Water–energy–food–climate nexus modelling	Integrated	Basin/regional	Multi-sectoral feedback included	UNESCO IHP DB	Asian river systems	Shrestha et al. (2017)
SWAT+ Human Module (USDA ARS 2019)	Human–environment feedback and adaptive management	Socio-hydrological	Basin/regional	Two-way socio-hydrological coupling	SWAT+ Portal	North America, Asia	Bieger et al. (2017), Arnold et al. (2022)
EcoHydro/ HYPE (SMHI 2008)	Ecohydrological linkages between biota and flow/ quality	Integrated	Catchment—national	Biotic feedback through nutrient cycling	SMHI HYPE DB	Scandinavia and Europe	Lindström et al. (2010)
HydroBID (IDB 2014)	Climate-informed hydrological simulations for Latin America	Hydrological + Policy	Basin /regional	Water allocation with basic quality parameters	Inter-American Development Bank	Andes and Caribbean basins	Valdés et al. (2016)

(continued)

Table 2.1 (continued)

Model and origin	Primary focus/process representation	Model category	Scale/spatial resolution	Integration of quantity and quality	Data source/ repository	Typical applications and basin examples	Validated reference
SPARROW (USGS 2003)	Spatially referenced regression for nutrient loading	Statistical –Empirical	Basin/regional	Regression-based flow–load relation	USGS SPARROW DB	Mississippi Basin	Smith et al. (1997)
OpenStreams/ OpenMI (European Consortium 2010)	Framework for linking different hydrological and WQ models	Meta-coupling platform	Multi-scale	Modular integration of existing models	OpenMI Association	EU rivers and lakes	Gregersen et al. (2007)
MODFLOW 6—SEAWAT (USGS 2017)	Density-dependent flow and solute transport in coastal aquifers	Coupled	3-D aquifer	Full salinity–flow interaction	USGS MODFLOW 6 Portal	Coastal delta regions	Langevin et al. (2017)
MIKE Integrated/ EcoLab (DHI 2016)	Ecosystem-based integrated modelling platform	Integrated	Multi-scale	Dynamic coupling among flow, biogeochemistry, and ecology	DHI EcoLab DB	Baltic Sea, Mekong Delta	DHI (2016)
PCR-GLOBWB (Delft Univ. 2011)	Global water balance and quality under climate change	Global Hydrological	0.5° grid global	Coupled flow–water use–nutrient export	Utrecht Univ. DB	Global assessments	van Beek et al. (2011)

beyond physical processes to incorporate human drivers, socio-economic feedbacks, and sustainability frameworks.

2.3.1 The Human–Water Nexus: Health, Food Security, and Livelihoods

Water scarcity and pollution directly impact human health and survival. Globally, more than two billion people still lack access to safely managed drinking water, and contaminated supplies contribute to diarrheal diseases that disproportionately affect children (WHO and UNICEF 2019). The interplay between water quantity and quality intensifies these risks: during droughts, limited flows reduce dilution of pathogens, heavy metals, and nutrients, raising exposure levels in drinking water sources (Whitehead et al. 2009; Wang et al. 2020). In agricultural landscapes, runoff following extreme rainfall events transports fertilizers and pesticides into rivers and reservoirs, creating toxic blooms that affect both rural and urban populations (Carpenter et al. 1998). Hence, variability in quantity directly amplifies threats to quality, with cascading human health consequences.

Food security is also intricately linked to water dynamics. Irrigation accounts for around 70% of global freshwater withdrawals, and both insufficient quantity and poor quality can undermine agricultural productivity (FAO 2017). Salinity, often worsened by excessive irrigation in arid climates, reduces soil fertility and crop yields, creating negative feedback loops between water management and agricultural sustainability (Qadir et al. 2014). Nutrient-rich runoff, while beneficial in small amounts, often crosses thresholds that trigger eutrophication and fish kills, thereby damaging fisheries that provide vital protein for many low-income communities (Smith et al. 2006). Drought-driven concentration of pollutants further limits safe irrigation options, forcing farmers to use contaminated water and incurring long-term health and ecological costs (Molden 2007). These interactions underscore the central role of water quality–quantity dynamics in sustaining food and livelihood systems.

Livelihoods in sectors such as agriculture, fisheries, energy, and tourism depend heavily on water services, and fluctuations in quality and quantity have direct economic impacts. Hydropower generation, for instance, is sensitive to both reduced river discharge and sedimentation caused by land-use change and extreme rainfall (Zarfl et al. 2015). Tourism in lakes and coastal areas suffers when algal blooms driven by nutrient-rich runoff reduce water quality and recreational value (Paerl and Otten 2013). Artisanal and subsistence fisheries collapse when drought-induced low flows coincide with oxygen depletion, depriving vulnerable populations of both food and income (Allan et al. 2005). Thus, integrating the human–water nexus into assessments of water quality–quantity interlinkages is essential to capture the full breadth of societal risks and opportunities. Figure 2.2 illustrates that QnQ interactions directly influence health risks, food security, livelihoods, and ecosystem condition, while human water demand and management decisions feed back to shape both availability

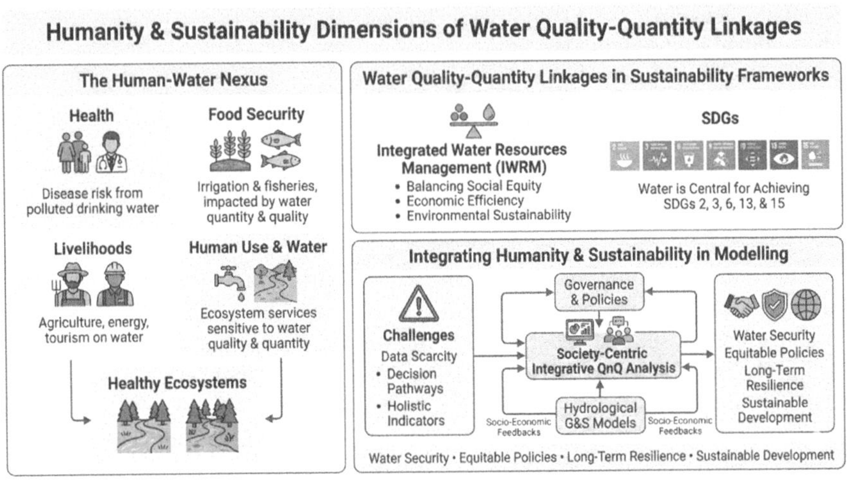

Fig. 2.2 Humanity and Sustainability Dimensions of Water Quality–Quantity Linkages. Conceptual framework linking water quality–quantity (QnQ) dynamics with human well-being and sustainability outcomes. It connects the human–water nexus, IWRM/SDG frameworks, and society-centric integrative modelling to support water security and equitable development

and pollution stress. It situates these couplings within sustainability frameworks such as IWRM and the SDGs, highlighting the need to manage both quality and quantity.

2.3.2 *Water Quality–Quantity Linkages in Sustainability Frameworks*

The imperative to sustain both water quality and quantity has been institutionalized in global frameworks, particularly Integrated Water Resources Management (IWRM) and the United Nations Sustainable Development Goals (SDGs). IWRM promotes coordinated development and management of water, land, and related resources, with explicit emphasis on balancing social equity, economic efficiency, and environmental sustainability. In practice, however, many IWRM implementations have struggled to incorporate robust science on water-quality–quantity linkages, often focusing more on allocation and supply while underestimating pollutant dynamics and ecological thresholds (Biswas 2004). Embedding coupled models that account for both hydrological variability and pollutant fate could substantially strengthen IWRM's analytical foundation, particularly for trade-off analyses between sectors and regions.

The SDGs reinforce the centrality of water for sustainability. SDG 6 calls for ensuring availability and sustainable management of water and sanitation for all, including explicit targets for both quality (reducing pollution, eliminating dumping, minimizing hazardous releases) and quantity (increasing water-use efficiency and sustainable withdrawals) (UN 2015). Beyond SDG 6, water also underpins SDG 2 (zero hunger), SDG 3 (good health and well-being), SDG 13 (climate action), and SDG 15 (life on land). Research shows that achieving water-related targets requires explicit recognition of the quality–quantity nexus: managing scarcity without controlling pollution risks undermining health and agriculture, while improving treatment without ensuring sufficient flows may leave ecosystems degraded (Vörösmarty et al. 2010; UNESCO 2020). For example, many river basins in Asia and Africa are projected to experience simultaneous increases in drought frequency and pollutant loading, highlighting the need for coordinated adaptation strategies (Kumar et al. 2020). Coupling water quantity and quality modelling within sustainability frameworks, therefore, provides a means to operationalize the SDGs and identify policy-relevant interventions.

The concept of water security also provides a unifying lens. Defined as the reliable availability of an acceptable quantity and quality of water for health, livelihoods, and production, coupled with an acceptable level of water-related risk, water security bridges scientific, policy, and societal concerns (Grey and Sadoff 2007; Agarwal et al. 2024). In this framing, water quality–quantity linkages become central: degraded quality reduces usable supply, while scarcity magnifies pollution risks. Water security assessments increasingly incorporate indices of both hydrological variability and pollutant stress, but these remain underutilized in guiding investment or governance reforms (Zeitoun et al. 2016). Strengthening their integration is thus critical for realizing sustainability commitments.

2.3.3 *Integrating Humanity and Sustainability into Modelling Frameworks*

Despite substantial advances in hydrological and water-quality modelling, the integration of **human and sustainability dimensions** remains only partially realized. Most existing frameworks still treat human activities—such as abstraction, land-use change, and reservoir operation—as boundary conditions or external forcings, rather than as **dynamic, interactive processes** within the system (Blöschl et al. 2013). This approach overlooks the crucial feedback loops through which human decisions alter hydrological and quality regimes, and these altered regimes, in turn, reshape future choices and adaptation. For instance, intensive irrigation not only reduces downstream flows but also increases salinity and nutrient runoff, directly influencing subsequent crop decisions and potentially generating transboundary conflicts over limited water resources (Qadir et al. 2014). Models that fail to internalize such socio-ecological feedback risk producing outputs that are scientifically detailed but detached from real-world governance and sustainability challenges.

Emerging approaches now aim to close this gap by coupling physical models with **socio-economic and behavioural modules**. Agent-Based Models (ABMs) simulate the micro-decisions of users such as farmers and industries, allowing exploration of how individual and collective behaviours influence pollution and water use, and how these, in turn, feed back into regulatory mechanisms (Schlüter et al. 2012). Integrated assessment models and socio-hydrological frameworks combine hydrology, economics, and ecology to evaluate multi-sectoral trade-offs—between agriculture, energy, and environment—under climate stress and competing demands (Biemans et al. 2019). Likewise, participatory modelling approaches that engage local stakeholders directly in scenario design improve social legitimacy and practical relevance, particularly when sustainability trade-offs are contested (Voinov and Bousquet 2010).

Integrating sustainability requires attention not only to physical processes but also to **equity and vulnerability**. Climate-induced water scarcity and pollution disproportionately affect marginalized groups that lack infrastructure and adaptive capacity (Sadoff et al. 2015). Modelling frameworks that average outcomes across regions risk masking these disparities and, consequently, reinforcing existing inequities. Sustainability-sensitive models must therefore incorporate **equity-based indicators**—for example, differential access to safe water, treatment costs, or the vulnerability of downstream ecosystems—to ensure that predictions inform fair and just policy responses (Zeitoun et al. 2016). In addition, uncertainty assessment should move beyond technical sensitivity analysis to include **social resilience**, examining how systems maintain core functions under multiple futures (Hallegatte et al. 2012).

Bridging the gap between modelling and sustainability action also demands **cross-disciplinary collaboration**. Hydrologists, economists, ecologists, and social scientists must work together to ensure that models reflect both process complexity and policy relevance. Linking physical and water-quality models with sustainability assessment tools such as **life-cycle analysis**, **ecosystem service valuation**, and **SDG indicators** offers a pragmatic path forward (Falkenmark and Rockström 2004). The inclusion of such approaches enables models to account not only for hydrological efficiency but also for human welfare, ecosystem thresholds, and long-term resilience.

To consolidate these ideas, the **Integrated Machine Learning–Humanity–Sustainability Framework for Coupled Water Quantity–Quality (QnQ) Systems** (Fig. 2.3) was developed to illustrate the sequential and feedback relationships between data inputs, modelling processes, human water use, and sustainability outcomes. The framework bridges physical science with human dimensions, showing how artificial intelligence and data analytics can enhance socio-hydrological understanding and support sustainable decision-making under increasing climate and anthropogenic pressures.

The framework illustrates the integration of multi-source hydrological, hydrochemical, and socio-economic data with advanced machine learning techniques to capture the interlinked behaviour of water quantity and quality. It shows how predictive analytics based on Random Forest, XGBoost, LSTM, and SVM models can identify nonlinear relationships between flow and pollution dynamics, translating them into human-relevant outcomes for agriculture, industry, and domestic sectors. The feedback chain culminates in **sustainability outcomes** aligned with **SDG 6 (Clean Water and Sanitation)**, ecosystem health, and equitable resource distribution. The directional flow—from data inputs to modelling core, QnQ interlinkages,

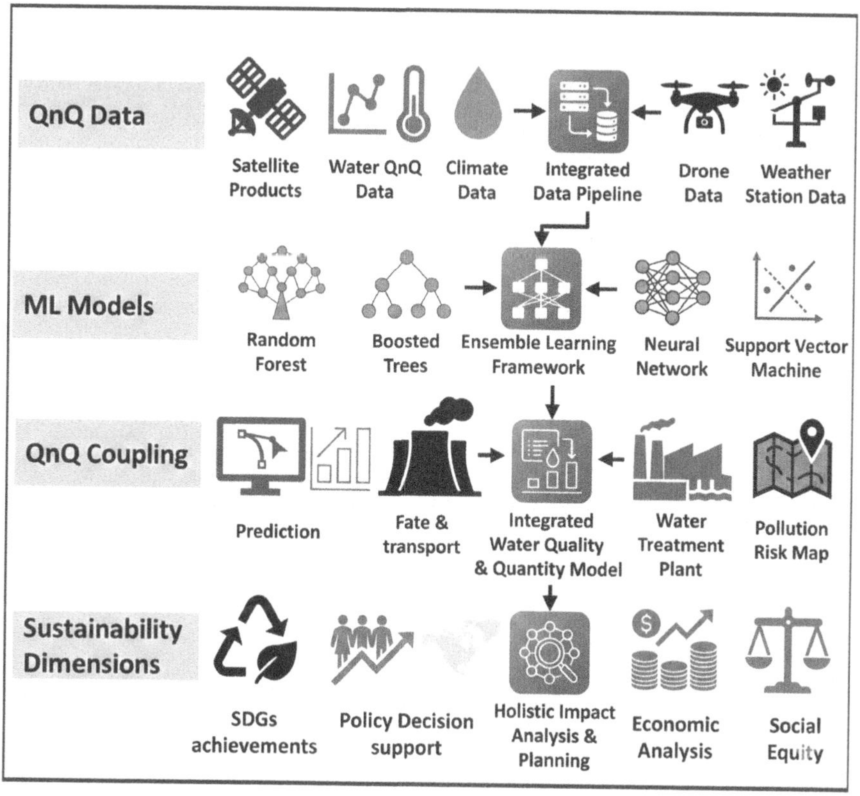

Fig. 2.3 Integrated Machine Learning–Humanity–Sustainability Framework for Coupled Water Quantity–Quality (QnQ) Systems. The framework illustrates how multi-source data and machine learning models (RF, XGBoost, LSTM, SVM) are linked to represent feedback between water quantity and quality. It connects hydrological and hydrochemical predictions with human water-use sectors and sustainability outcomes such as SDG 6, ecosystem health, and water security, supporting adaptive and equitable water governance

human water-use sectors, and sustainability layers—emphasizes the cyclic, adaptive nature of water governance informed by machine learning. The framework thus provides a **human-centred, sustainability-driven pathway** for integrating artificial intelligence into water-resource planning and management.

2.4 Future Directions

Water quality–quantity modelling is moving toward several priorities. First, fully integrated frameworks are needed to represent feedback among hydrological variability, pollutant transport, and human water use. Although SWAT and MIKE SHE couple physical and biogeochemical processes, future models must treat human activities as dynamic drivers (e.g., land-use change, irrigation withdrawals, reservoir operations, and pollution control) rather than fixed boundary conditions (Blöschl et al. 2013; Schlüter et al. 2012). Second, climate indices and extremes should be embedded more systematically. SPI is useful for drought–wet characterization, but coupling it with SPEI and PDSI can better represent evaporative demand and soil moisture balance (Vicente-Serrano et al. 2010; Dai 2011), and hybrid climate-index + machine-learning approaches can improve forecasting under non-stationarity (Mosavi et al. 2018).

Third, improved data availability is essential: remote sensing can track water-quality proxies (chlorophyll a, turbidity, suspended sediments) at broad scales (Duan et al. 2019), while open data and citizen science can reduce monitoring gaps and strengthen local relevance (Buytaert et al. 2014). Fourth, models must better connect to sustainability and equity by embedding SDG targets, ecosystem-service valuation, and equity-sensitive metrics in outputs (Falkenmark and Rockström 2004; UNESCO 2020), supported by participatory modelling to ensure usability and legitimacy (Voinov and Bousquet 2010). Finally, uncertainty and resilience must be central: robust decision-making, scenarios, and resilience assessments can shift modelling from deterministic prediction to adaptive pathways under deep uncertainty (Hallegatte et al. 2012).

2.5 Conclusion

The present review reveals that global water security requires an understanding of the dynamic linkages between water quantity and quality. While hydrological and hydrochemical models have advanced process representation, their largely separate development limits the capture of coupled feedbacks. The study reflects a shift toward integrated platforms that simulate hydrological–biogeochemical interactions, yet data scarcity, uncertainty propagation, and weak treatment of hydroclimatic variability still constrain applicability. The nitrate–SPI-9 example illustrates how water quality responds to hydroclimatic fluctuations, emphasizing that scarcity and contamination arise from shared climatic and anthropogenic pressures and that climate indices strengthen empirical coupling of hydrological and chemical datasets for identifying drought–pollution hotspots. A major advance is machine-learning-assisted modelling, where Random Forests, XGBoost, Support Vector Machines, and LSTMs can assimilate diverse data and learn nonlinear climatic–hydrological–chemical dependencies, thereby improving predictive skill and supporting spatio-temporal complexity. Embedded within a human- and sustainability-centred approach, these tools function as decision-support systems for adaptive management.

References

Agarwal V, Kumar M, Panday DP, Zang J, Munoz-Arriola F (2024) Unlocking the potential of remote sensing for arsenic contamination detection and management: challenges and perspectives. Curr Opin Environ Sci Health 100578. https://doi.org/10.1016/j.coesh.2024.100578

Allan JD, Abell R, Hogan Z, Revenga C, Taylor BW, Welcomme RL, Winemiller K (2005) Overfishing of inland waters. Bioscience 55(12):1041–1051

Ambrose RB, Wool TA, Martin JL (1993) The water quality analysis simulation program, WASP5 part B: the WASP5 input dataset. Environmental Research Laboratory, US Environmental Protection Agency, Athens, GA

Arnold JG, Moriasi DN, Gassman PW, Abbaspour KC, White MJ, Srinivasan R, Jha MK (2012) SWAT: model use, calibration, and validation. Trans ASABE 55(4):1491–1508

Beven K (2012) Rainfall–runoff modelling: the primer, 2nd edn. Wiley-Blackwell

Beven KJ, Kirkby MJ (1979) A physically-based variable contributing area model of basin hydrology: hydrological Science Bulletin, vol 24

Bicknell BR, Donigian Jr AS, Barnwell TA (1985) Modeling water quality and the effects of agricultural best management practices in the Iowa River Basin. Water Sci Technol 17(6–7):1141–1153

Biemans H, Siderius C, Mishra A, Ahmad B (2019) Crop production losses due to river floods. Environ Res Lett 14(1):014009. https://doi.org/10.1088/1748-9326/aaeb83

Biswas AK (2004) Integrated water resources management: a reassessment. Water Int 29(2):248–256. https://doi.org/10.1080/02508060408691775

Blöschl G, Sivapalan M, Wagener T, Viglione A, Savenije H (eds) (2013) Runoff prediction in ungauged basins: synthesis across processes, places, and scales. Cambridge University Press

Bonsor HC, Shamsudduha M, Marchant BP, Macdonald AM, Taylor RG (2018) Seasonal and decadal groundwater changes in African sedimentary aquifers estimated using GRACE products and LSMs. Remote Sens 10(6):904

Bretas F, Casanova G, Crisman TL, Irujo AE, Martin L, Miralles-Wilhelm F, Castillo RM (2020) Water for the Future: water Security in Latin America and the Caribbean Strategy and Work Plan for the Inter-American Development Bank

Brown LC, Barnell TO (1987) The enhanced stream water quality QUAL2E and QUAL2EUNCAS: documentation and user manual. Report EPA, Athens, GA, USA, Vol 600, pp 3–87

Buytaert W, Zulkafli Z, Grainger S, Acosta L, Alemie TC, Bastiaensen J, Porras I (2014) Citizen science in hydrology and water resources: opportunities for knowledge generation, ecosystem service management, and sustainable development. Front Earth Sci 2:26. https://doi.org/10.3389/feart.2014.00026

Carpenter SR, Caraco NF, Correll DL, Howarth RW, Sharpley AN, Smith VH (1998) Non-point pollution of surface waters with phosphorus and nitrogen. Ecol Appl 8(3):559–568

Chhipi-Shrestha G, Hewage K, Sadiq R (2017) Water–energy–carbon nexus modeling for urban water systems: system dynamics approach. J Water Resour Plan Manag 143(6):04017016. https://doi.org/10.1061/(ASCE)WR.1943-5452.0000765

Cole TM, Wells SA (2006) CE-QUAL-W2: a two-dimensional, laterally averaged, hydrodynamic and water quality model, version 3.5. https://www.cee.pdx.edu/w2/download_model_information.html

Dai A (2011) Drought under global warming: a review. Wiley Interdisc Rev Clim Change 2(1):45–65. https://doi.org/10.1002/wcc.81

DHI (2016) MIKE Powered by DHI: ecoLab user guide. Danish Hydraulic Institute. https://www.mikepoweredbydhi.com

Döll P, Kaspar F, Lehner B (2003) A global hydrological model for deriving water availability indicators: model tuning and validation. J Hydrol 270(1–2):105–134. https://doi.org/10.1016/S0022-1694(02)00283-4

Duan H, Ma R, Xu J, Zhang Y, Zhang B, Song K (2019) Remote sensing of water quality in lakes and rivers. Remote Sens Environ 221:58–73. https://doi.org/10.1016/j.rse.2018.11.008

FAO (2017) The future of food and agriculture: trends and challenges. Food and Agriculture Organization of the United Nations, Rome
Falkenmark M, Rockström J (2004) Balancing water for humans and nature: the new approach in ecohydrology. Earthscan
Farooq M, Thulasiraman AV, Manzoor Z, Tripathi S, Nisa FU, Farooq A, Gani KM (2024) Comprehensive characterization of unscientifically disposed municipal solid waste (MSW) in the Kashmir Region, India. Environ Monitor Assess 196(5):450
Grey D, Sadoff CW (2007) Sink or swim? Water security for growth and development. Water Policy 9(6):545–571. https://doi.org/10.2166/wp.2007.021
Graham DN, Butts MB (2005) Flexible, integrated watershed modelling with MIKE SHE. Watershed models, 849336090:245–272.
Gregersen JB, Gijsbers PJA, Westen SJP (2007) OpenMI: open modelling interface. J Hydroinform 9(3):175–191
Hallegatte S, Shah A, Lempert R, Brown C, Gill S (2012) Investment decision making under deep uncertainty. World Bank Policy Research Working Paper 6193. https://doi.org/10.1596/1813-9450-6193
Kaushal SS, Gold AJ, Bernal S, Tank JL (2018) Diverse water quality responses to extreme climate events: an introduction. Biogeochemistry 141(3):273–279
Kumar P, Singh R, Singh VP (2020) Water quality and drought linkages in India. Environ Monit Assess 192:321. https://doi.org/10.1007/s10661-020-8233-5
Kumar M, Saxena A, Tripathi S, Panday DP, Torres-Martínez JA (2025a) Harnessing microplastic contaminant hydro (bio)logy: Proxies for flood impact, surface-groundwater connectivity, and urbanisation. Curr Opin Environ Sci Health 100696. https://doi.org/10.1016/j.coesh.2025.100696
Kumar M, Panday DP, Goswami R, Neog N, Bahukhandi KD, Zhang J, Bhattacharya P (2025b) Dynamics of arsenic co-contamination in the groundwater of the Eastern Himalayas: the biogeochemical drivers and GRACE-based evidence. ACS ES&T Water. https://doi.org/10.1021/acsestwater.4c00106
Kumar M, Tripathi S, Singh R, Rai SP, Das N, Agarwal V, Torres-Martinez JA (2026a) Entropy and seasonal isotopic duality reveal the sustainability paradox of the upper Ganga River. Scientific Reports. https://doi.org/10.1038/s41598-026-44251-8
Kumar M, Saxena A, Tripathi S, Panday DP, Das N, Singh R, Mahlknecht J (2026b) Localized or diffusive: insights into the source and geochemical dynamics of dissolved mercury contamination in groundwater of the mid-Gangetic Plains. npj Clean Water. https://doi.org/10.1038/s41545-026-00566-7
Langevin CD, Hughes JD, Banta ER, Niswonger RG, Panday S, Provost AM (2017) Documentation for the MODFLOW 6 groundwater flow model (No. 6-A55). US Geological Survey
Lindström G, Johansson B, Persson M, Gardelin M, Bergström S (1997) Development and test of the distributed HBV-96 hydrological model. J Hydrol 201(1–4):272–288. https://doi.org/10.1016/S0022-1694(97)00041-3
Lindström G, Pers C, Rosberg J, Strömqvist J, Arheimer B (2010) Development and testing of the HYPE (Hydrological Predictions for the Environment) water quality model for different spatial scales. Hydrol Res 41(3–4):295–319. https://doi.org/10.2166/nh.2010.007
McKee TB, Doesken NJ, Kleist J (1993) The relationship of drought frequency and duration to time scales. In: Proceedings of the 8th conference on applied climatology, vol 17, no 22, pp 179–183
Molden D (ed) (2007) Water for food, water for life: a comprehensive assessment of water management in agriculture. Earthscan/IWMI
Mosavi A, Ozturk P, Chau KW (2018) Flood prediction using machine learning models: literature review. Water 10(11):1536. https://doi.org/10.3390/w10111536
Oelsner GP, Stets EG (2019) Recent trends in nutrient and sediment loading to coastal areas of the conterminous US: insights and global context. Sci Total Environ 654:1225–1240

Panday DP, Kumari A, Kumar M (2025) Alkalinity-salinity-sustainability: decadal groundwater trends and its impact on agricultural water quality in the Indian Peninsula. Sci Total Environ 978:179459. https://doi.org/10.1016/j.scitotenv.2025.179459

Panday DP, Kumar M, Agarwal V, Torres-Martínez JA, Mahlknecht J (2024) Corroboration of arsenic variation over the Indian Peninsula through standardized precipitation evapotranspiration indices and groundwater level fluctuations: water quantity indicators for water quality prediction. Sci Total Environ 176339. https://doi.org/10.1016/j.scitotenv.2024.176339

Panday DP, Kumar M (2024) Decoding uranium variation over the Indian Peninsula through leveraging standardized precipitation evapotranspiration indices and groundwater level fluctuations. Groundw Sustain Dev 101241. https://doi.org/10.1016/J.GSD.2024.101241

Paerl HW, Otten TG (2013) Harmful cyanobacterial blooms: causes, consequences, and controls. Microb Ecol 65(4):995–1010. https://doi.org/10.1007/s00248-012-0159-y

Qadir M, Quillérou E, Nangia V, Murtaza G, Singh M, Thomas RJ, Noble AD (2014) Economics of salt-induced land degradation and restoration. Nat Res Forum 38(4):282–295. https://doi.org/10.1111/1477-8947.12054

Rossman LA (2010) Storm water management model user's manual

Sadoff CW, Hall JW, Grey D, Aerts JCJH, Ait-Kadi M, Brown C, Wiberg D (2015) Securing water, sustaining growth: report of the GWP/OECD task force on water security and sustainable growth. University of Oxford

Scanlon BR, Jolly I, Sophocleous M, Zhang L (2007) Global impacts of conversions from natural to agricultural ecosystems on water resources: quantity versus quality. Water Resour Res 43(3)

Scanlon BR, Faunt CC, Longuevergne L, Reedy RC, Alley WM, McGuire VL, McMahon PB (2012) Groundwater depletion and sustainability of irrigation in the US high plains and central valley. Proc Natl Acad Sci 109(24):9320–9325

Schlüter M, Müller B, Frank K (2012) New horizons for managing the environment: a review of coupled social–ecological systems modelling. Nat Resour Model 25(1):219–272. https://doi.org/10.1111/j.1939-7445.2011.00108.x

Smith RA, Schwarz GE, Alexander RB (1997) Regional interpretation of water-quality monitoring data. Water Resour Res 33(12):2781–2798. https://doi.org/10.1029/97WR02171

Smith VH, Tilman GD, Nekola JC (2006) Eutrophication: Impacts of excess nutrient inputs on freshwater, marine, and terrestrial ecosystems. Environ Pollut 100(1–3):179–196. https://doi.org/10.1016/j.envpol.2005.07.023

Streeter HW, Phelps EB (1925) A study of the pollution and natural purification of the Ohio River (No. 146). United States Public Health Service

Sutadian AD, Muttil N, Yilmaz AG, Perera BJC (2016) Development of river water quality indices—a review. Environ Monit Assess 188(1):58

Taylor RG, Scanlon B, Döll P, Rodell M, Van Beek R, Wada Y, Treidel H (2013) Ground water and climate change. Nat Clim Chang 3(4):322–329

Tripathi PM, Agarwal V, Mishra V, Singh BVR (2025) Options and strategies for managing droughts: case studies from agro-ecological regions. In Water Sustainability and Hydrological Extremes (pp 315–331). Elsevier. https://doi.org/10.1016/B978-0-443-21499-8.00016-7

UN (2015) Transforming our world: the 2030 agenda for sustainable development. United Nations, New York

UNESCO (2020) The United Nations World Water Development Report 2020: water and climate change. UNESCO Publishing, Paris

Van Beek LPH, Wada Y, Bierkens MF (2011) Global monthly water stress: 1. Water balance and water availability. Water Resour Res 47(7)

Vicente-Serrano SM, Beguería S, López-Moreno JI (2010) A multi-scalar drought index sensitive to global warming: the SPEI. J Clim 23(7):1696–1718. https://doi.org/10.1175/2009JCLI2909.1

Voinov A, Bousquet F (2010) Modelling with stakeholders. Environ Model Softw 25(11):1268–1281. https://doi.org/10.1016/j.envsoft.2010.03.007

Vörösmarty CJ, McIntyre PB, Gessner MO, Dudgeon D, Prusevich A, Green P, Davies PM (2010) Global threats to human water security and river biodiversity. Nature 467(7315):555–561. https://doi.org/10.1038/nature09440

Wang J, Jin J, Liu J, Lu C (2020) Drought impacts on river water quality: a global review. Water 12(6):1612. https://doi.org/10.3390/w12061612

Whitehead PG, Wilson EJ, Butterfield D (2009) A semi-distributed integrated flow and water quality model (INCA). Hydrol Earth Syst Sci 2(2–3):559–582. https://doi.org/10.5194/hessd-2-559-2005

WHO & UNICEF (2019) Progress on household drinking water, sanitation and hygiene 2000–2017: special focus on inequalities. World Health Organization (WHO) and UNICEF, Geneva

Yates D, Sieber J, Purkey D, Huber-Lee A (2005) WEAP21—A demand-, priority-, and preference-driven water planning model: part 1: model characteristics. Water Int 30(4):487–500. https://doi.org/10.1080/02508060508691893

Zarfl C, Lumsdon AE, Berlekamp J, Tydecks L, Tockner K (2015) A global boom in hydropower dam construction. Aquat Sci 77(1):161–170

Zeitoun M, Lankford B, Bakker K (2016) Managing water scarcity: alternative frameworks and perspectives. Routledge

Zhang H, Cheng H, Cai Y, Spötl C, Kathayat G, Sinha A, Tan L (2018) Hydroclimatic variations in southeastern China during the 4.2 ka event reflected by stalagmite records. Clim Past 14(11):1805–1817

Zheng C, Wang PP (1999) MT3DMS: a modular three-dimensional multispecies transport model. Strategic Environmental Research and Development Program, 1–40

Chapter 3
Predictive Modelling for Contaminant Forecasting

Abstract This chapter examines predictive modelling approaches for contaminant forecasting in water systems during the Anthropocene era. As water bodies face increasing pressures from diverse pollutants, including pharmaceuticals, microplastics, and per- and polyfluoroalkyl substances (PFAS), predictive models have become essential tools for proactive water quality management. The chapter establishes a theoretical framework for contaminant transport, covering fundamental processes such as advection, dispersion, and chemical transformation that govern pollutant movement and fate in aquatic environments. It explores the full spectrum of modelling approaches, from traditional process-based mechanistic models to data-driven machine learning techniques, discussing their relative strengths and appropriate application contexts. Attention is given to hybrid frameworks that combine physical understanding with pattern-recognition capabilities to enhance predictive performance. Practical implementation aspects are addressed, including data requirements, calibration procedures, sensitivity analysis, and uncertainty quantification essential for reliable forecasting. The chapter examines integration with decision support systems to translate predictions into actionable management strategies, and discusses emerging trends including artificial intelligence, digital twin technology, and real-time sensor networks. By bridging theoretical advances with practical applications, this chapter guides the selection, implementation, and evaluation of predictive models to address contamination challenges and support resilient, sustainable water resource management under conditions of unprecedented environmental change.

Keywords Contaminant transport modelling · Water quality forecasting · Machine learning · Advection–dispersion · Uncertainty quantification · Decision support systems

M. Kumar et al., *Bridging Water Quality and Quantity in the Anthropocene*, SpringerBriefs in Water Science and Technology,
https://doi.org/10.1007/978-3-032-25655-3_3

3.1 Introduction

The Anthropocene epoch has fundamentally transformed global water systems through unprecedented human activities, creating complex contamination challenges that threaten human health and ecosystem integrity. Contemporary water bodies face mounting pressures from diverse pollutants, including legacy industrial chemicals and emerging contaminants such as pharmaceuticals, microplastics, and per- and polyfluoroalkyl substances (PFAS). These novel pollutants exhibit unique transport behaviours and transformation pathways that conventional monitoring approaches struggle to address. Combined with climate-driven hydrological changes, the synergistic effects of multiple contaminants create challenges for prediction, demanding innovative management strategies (Singh et al. 2024).

Predictive modelling for contaminant forecasting has emerged as an essential tool bridging environmental monitoring and effective water quality management (Miller et al. 2025; Mathur et al. 2025). Unlike reactive approaches that respond post-detection, predictive models enable managers to anticipate pollution events, evaluate intervention strategies, and implement proactive mitigation before irreversible damage occurs. This capability is invaluable for emergency spill response, network monitoring optimization, treatment system design, and regulatory framework development. By providing quantitative forecasts of water quality conditions, these models support risk-based decision-making and optimize resource allocation in increasingly constrained management environments (Haghighat et al. 2026; van der Heijden et al. 2025).

The evolution of contamination forecasting reflects broader technological advances, with traditional process-based models now complemented by data-driven approaches leveraging machine learning and artificial intelligence. This convergence has spawned hybrid frameworks that combine mechanistic understanding with pattern-recognition capabilities, offering superior predictive performance across diverse scenarios. Real-time sensor networks and remote sensing technologies have further enhanced modelling capabilities, enabling near-instantaneous contamination warnings (Agarwal et al. 2021, 2020; Qin et al. 2021; Mishra et al. 2024). However, implementation faces persistent challenges, including data scarcity in developing regions, the computational complexity of spatiotemporal dependencies, and requirements for uncertainty quantification. Model transferability across different systems remains problematic, often requiring extensive recalibration.

Selecting appropriate modelling approaches depends on multiple factors: contaminant characteristics, spatiotemporal scales, data availability, computational resources, and management objectives (Talukdar et al. 2023; Ciampi et al. 2022; Banerjee et al. 2023; Panday and Kumar 2024). Process-based models excel when transport mechanisms are well understood, and sufficient parameterization data are available, providing mechanistic insights and scenario analysis beyond historical observations. Data-driven models prove advantageous for complex non-linear relationships or when rapid operational predictions are required. Emerging hybrid approaches leverage both paradigms' strengths, using physical constraints to guide machine learning while maintaining interpretability. Climate change adds complexity through altered hydrological regimes, temperature-dependent reaction rates, and extreme events that create episodic pollution releases, requiring adaptive modelling frameworks that incorporate non-stationary conditions.

Throughout this chapter, emphasis is placed on bridging theoretical advances with practical applications, recognizing that effective water quality management requires not only accurate predictions but also appropriate communication of uncertainties to stakeholders. By providing a balanced treatment of established and emerging approaches, this chapter equips readers to select, implement, and evaluate predictive models appropriate to their specific contamination challenges, ultimately contributing to more resilient and sustainable water systems in an era of unprecedented environmental change.

3.2 Theoretical Framework for Contaminant Transport

Understanding the fundamental processes governing contaminant transport in aquatic systems is essential for developing accurate predictive models. The movement and transformation of pollutants through water bodies involve complex interactions between physical, chemical, and biological processes that operate across multiple spatial and temporal scales. This section presents the theoretical framework underlying contaminant fate and transport, establishing the mathematical foundations and conceptual understanding necessary for effective model development and application.

3.2.1 Fundamental Transport Processes

Contaminant transport in water systems is primarily governed by advection and dispersion mechanisms, which determine the spatial distribution and temporal evolution of pollutant concentrations (Fig. 3.1). Advection represents the bulk movement of contaminants with flowing water, driven by velocity fields that vary spatially and temporally in response to hydraulic gradients, channel geometry, and boundary conditions (Zhang et al. 2023; Saadat et al. 2024). In rivers and streams, advective transport typically dominates, with contaminants moving downstream at velocities approximating the mean flow velocity. However, in lakes, estuaries, and groundwater systems, advection patterns become more complex, influenced by density stratification, wind-driven circulation, and tidal forces that create three-dimensional flow fields requiring sophisticated mathematical representation.

Dispersion encompasses the spreading and mixing of contaminants due to velocity variations at multiple scales (Lin et al. 2024). Molecular diffusion, though typically negligible in surface waters except near boundaries, provides the fundamental mixing mechanism in stagnant conditions. Turbulent diffusion, driven by chaotic

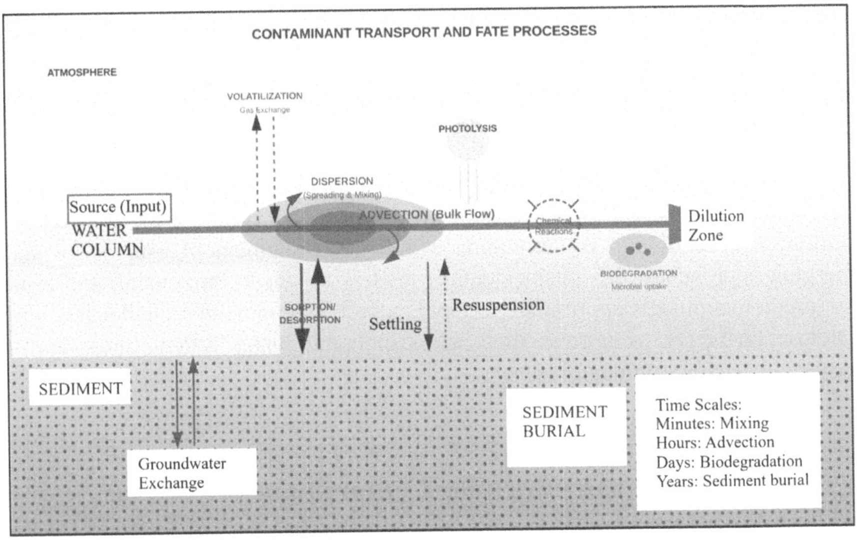

Fig. 3.1 Conceptual diagram of contaminant transport and fate processes in aquatic systems. The figure illustrates the major physical, chemical, and biological mechanisms affecting pollutant behaviour across three environmental compartments: atmosphere, water column, and sediment

fluid motions, dominates mixing in most natural water bodies, yielding effective dispersion coefficients that are orders of magnitude larger than molecular values. The longitudinal dispersion coefficient, particularly critical in river systems, results from the combined effects of vertical and lateral velocity gradients and captures the stretching and mixing of contaminant plumes as they move downstream.

Chemical transformation processes fundamentally alter contaminant characteristics and concentrations during transport. Hydrolysis reactions, particularly important for organic compounds, involve the breaking of chemical bonds through interaction with water molecules, often resulting in the formation of daughter products with different toxicological properties and transport behaviours. Photolysis, driven by solar radiation penetrating the water column, degrades photosensitive compounds, with reaction rates dependent on light intensity, water clarity, and compound-specific quantum yields. Oxidation–reduction reactions, mediated by dissolved oxygen levels and redox potential, control the speciation and mobility of many inorganic contaminants, particularly heavy metals whose oxidation states determine their solubility and toxicity (Linnik et al. 2023).

Sorption processes, encompassing both adsorption to particle surfaces and absorption into organic matter, significantly influence contaminant mobility and bioavailability (Sigmund et al. 2022; Masinga et al. 2024). The partitioning of pollutants between dissolved and particulate phases depends on compound hydrophobicity, sediment characteristics, and water chemistry parameters, including pH, ionic strength, and dissolved organic carbon concentrations (Liu et al. 2023). Linear equilibrium partitioning, commonly described through distribution coefficients, provides a first-order approximation suitable for many modelling applications. However, nonlinear sorption isotherms, such as the Freundlich or Langmuir relationships, become necessary when dealing with high contaminant concentrations or competitive sorption scenarios (Amrutha et al. 2023). Kinetic limitations to sorption, particularly during rapid transport events, require time-dependent formulations that account for mass-transfer resistances and intra-particle diffusion.

Biological transformation is a critical removal mechanism for many organic contaminants, as microorganisms metabolize pollutants as carbon and energy sources. Biodegradation rates depend on multiple factors, including microbial population density and diversity, nutrient availability, temperature, pH, and the presence of co-substrates or inhibitory compounds. First-order decay models, while simplistic, often adequately represent biodegradation in well-mixed systems with stable environmental conditions. However, Monod-type kinetics, incorporating substrate limitation and microbial growth dynamics, become necessary when modelling systems with variable contaminant loads or engineered bioremediation scenarios. The formation of metabolites during biodegradation introduces additional complexity, as these transformation products may exhibit different transport properties and toxicological impacts than parent compounds.

3.2.2 *Mathematical Foundations*

The mathematical description of contaminant transport in water systems builds upon fundamental fluid mechanics principles that most engineers and scientists encounter in their basic training. By starting with familiar equations like continuity, Bernoulli's, and Darcy's law, we can progressively develop the mathematical framework for contaminant transport modelling. This approach provides an intuitive understanding of how pollutants move through water bodies while establishing the quantitative tools necessary for predictive modelling.

Continuity Equation: The Foundation of Mass Conservation

The continuity equation (Eq. 3.1), expressing mass conservation in fluid flow, provides the starting point for understanding contaminant transport. In its simplest form for incompressible flow, the continuity equation states:

$$Q = A_1V_1 = A_2V_2 \tag{3.1}$$

where Q is the discharge, A is the cross-sectional area, and V is the velocity. This fundamental relationship states that in a pipe or channel with varying cross-sections, water speeds up as the area decreases and slows down as the area increases. For contaminant transport, this principle directly impacts how pollutants move through constrictions, expansions, and confluences in river systems (Seyedpour et al. 2023). Extending this to include contaminant mass, we apply the same conservation principle (Eqs. 3.2 and 3.3). The mass of contaminant entering a water body segment must equal the mass leaving plus any accumulation or loss within the segment:

$$\text{Accumulation} = \text{Input} - \text{Output} \pm \text{Reactions} \tag{3.2}$$

$$\frac{\partial(VC)}{\partial t} = \mathrm{Q_{in}} \cdot \mathrm{C_{in}} - \mathrm{Q_{out}} \cdot \mathrm{C_{out}} \pm \text{Reactions} \tag{3.3}$$

where V is the volume of water, C is the concentration, and Q represents flow rates. This simple mass balance serves as the basis for box models widely used in lake and reservoir contamination studies, in which water bodies are divided into well-mixed compartments.

From Darcy's Law to Groundwater Contaminant Transport

In groundwater contamination, Darcy's law provides the fundamental relationship for velocity. Originally developed for flow through sand filters, Darcy's law states (3.4):

$$\mathrm{v} = -\mathrm{K}\frac{dh}{dx} \tag{3.4}$$

where v is the specific discharge (Darcy velocity), K is the hydraulic conductivity, and dh/dx is the hydraulic gradient. The negative sign indicates flow from high to low hydraulic head. The actual pore water velocity, which determines contaminant transport speed, is:

$$\mathrm{v_x} = \frac{v}{n} = -\frac{K}{n} \cdot \frac{dh}{dx} \tag{3.5}$$

where n is the effective porosity. This relationship shows that contaminants move faster through materials with lower porosity, even when the Darcy velocity is the same. Combining Darcy's law with mass conservation for a contaminant yields the basic groundwater transport equation (Eq. 3.6):

$$\frac{\partial(nc)}{\partial t} = -\frac{\partial(v_x c)}{\partial x} + \frac{\partial}{\partial x}\left(nD\frac{\partial C}{\partial x}\right) - \lambda nC \tag{3.6}$$

This equation shows that contaminant concentration changes due to three processes: advection (first term on right), dispersion (second term), and decay (third term). The advection term directly incorporates the Darcy velocity, linking hydraulic conditions to contaminant movement.

Bernoulli's Principle and Surface Water Transport

While Bernoulli's equation primarily describes energy conservation in flowing fluids, its principles help understand contaminant behaviour in surface waters. The simplified Bernoulli equation along a streamline (Eq. 3.7):

$$P/\rho g + v^2/2g + z = \text{constant} \tag{3.7}$$

demonstrates that changes in velocity, pressure, and elevation are interconnected. In practical terms, this means that at constrictions where velocity increases (such as river rapids or narrow channels), the increased turbulence enhances the mixing and dispersion of contaminants. Conversely, in slow-moving pools where velocity decreases, contaminants may accumulate and undergo more complete reactions.

Advection–Dispersion Equation: Putting It All Together

Building on these fundamental principles, we arrive at the one-dimensional advection–dispersion equation, the workhorse of contaminant transport modelling (Eq. 3.8):

$$\partial C/\partial t + U\partial C/\partial x = D\partial^2 C/\partial x^2 - kC + S \tag{3.8}$$

Let's understand each term through our fundamental equations:

$\partial C/\partial t$	Change in concentration over time (from continuity principle)
$U\partial C/\partial x$	Advection term, where U is the average velocity (from continuity equation $Q = UA$)
$D\partial^2 C/\partial x^2$	Dispersion term, representing spreading due to velocity variations
kC	First-order decay (reaction term)
S	Source/sink term

The advection term moves the contaminant downstream at velocity U, directly related to discharge through $Q = UA$. The dispersion coefficient D captures all spreading processes, from molecular diffusion to turbulent mixing, as predicted by Bernoulli's principle at flow transitions.

The elegance of the advection–dispersion equation lies in its ability to capture complex transport phenomena through a relatively simple mathematical framework (Zhang et al. 2021; Sanskrityayn et al. 2021; Carrera et al. 2022). Each term has a clear physical meaning and can be estimated from measurable quantities. The advection velocity (U) can be determined from flow measurements using standard stream-gauging techniques, while the dispersion coefficient (D) can be estimated from channel geometry and flow characteristics, or determined through tracer studies. The decay rate (k) can be measured in laboratory experiments or estimated from field observations of contaminant attenuation. This direct connection between mathematical parameters and physical measurements makes the equation particularly valuable for practical applications, allowing engineers and scientists to predict contaminant concentrations without requiring detailed knowledge of every turbulent eddy or chemical interaction occurring in the water body.

The advection–dispersion equation also serves as a foundation that can be extended to address more complex scenarios. For reactive contaminants, additional terms can be added to represent specific transformation processes such as volatilization, photolysis, or biodegradation. For systems with significant sediment interaction, the equation can be modified to include sorption–desorption kinetics and bed-water exchange processes. In multidimensional applications, the equation naturally extends to include transport in the lateral and vertical directions, though this substantially increases computational requirements. The modular nature of the equation allows modelers to start with simple representations and progressively add complexity as data availability and management needs dictate, making it an invaluable tool for both screening-level assessments and detailed contamination studies.

Understanding the mathematical foundations rooted in these familiar engineering principles provides the necessary framework for selecting and applying appropriate modelling approaches discussed in subsequent sections. Whether employing sophisticated numerical models or simplified analytical solutions, the fundamental relationships between flow, transport, and transformation embodied in the advection–dispersion equation remain central to predicting contaminant fate. This mathematical understanding, combined with knowledge of system-specific characteristics and careful consideration of model assumptions and limitations, enables the development of reliable predictive tools essential to protecting water resources amid increasing contamination pressures. The transition from basic fluid mechanics principles to practical contaminant transport models demonstrates that complex environmental phenomena can be understood and predicted through the systematic application of fundamental scientific principles.

3.3 Predictive Modelling Approaches

Selecting appropriate modelling approaches for contaminant forecasting significantly influences prediction accuracy, computational requirements, and management utility (Fig. 3.2). The predictive modelling toolkit spans mechanistic process-based models rooted in physical principles to data-driven approaches leveraging machine learning, with empirical and hybrid methods occupying intermediate positions. Each paradigm offers distinct advantages, and optimal selection depends on data availability, system complexity, prediction timescales, and management objectives.

3.3.1 Data-Driven Models

Data-driven models identify patterns directly from observational data without explicit representation of underlying physical processes. These approaches prove particularly valuable for real-time predictions and complex non-linear relationships. Machine learning algorithms capture complex non-linear relationships without explicit functional forms. Random Forest models demonstrate strong performance through ensemble learning while providing variable importance measures. Support Vector Machines excel at classification tasks such as exceedance prediction. Artificial Neural Networks provide universal function approximation, while LSTM architectures handle sequential dependencies effectively. Deep learning approaches, including

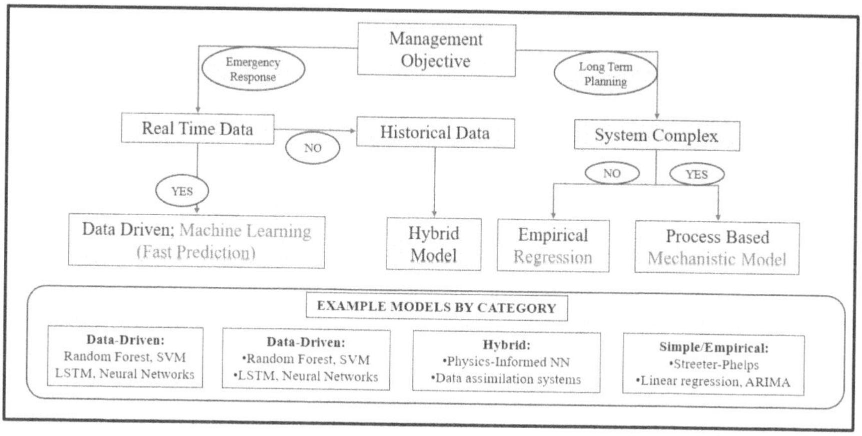

Fig. 3.2 Decision flowchart for selecting appropriate contamination forecasting modelling approaches based on management objectives and data availability

Convolutional Neural Networks, are well-suited for spatial pattern recognition, and Graph Neural Networks show promise for networked systems.

Data-driven models offer significant advantages: minimal process assumptions enable application to poorly understood systems; efficient handling of high-dimensional data; rapid prediction for real-time applications; and flexibility incorporating diverse data types. However, their empirical nature limits extrapolation beyond the training range, complex models' "black box" nature hampers interpretation, and performance critically depends on data quality. Lack of physical constraints may produce unrealistic predictions violating conservation laws.

3.3.2 *Hybrid Approaches*

Hybrid modelling approaches combine process-based and data-driven methods, leveraging mechanistic understanding while maintaining flexibility to capture empirical patterns (Zhou et al. 2021; Kumar and Nehdi 2026). These integrated frameworks address limitations of purely mechanistic or empirical approaches. Physics-informed neural networks incorporate physical laws directly into neural network training by including governing equations as constraints. In contaminant transport, the advection–dispersion equation serves as a soft constraint, with networks learning solutions and unknown parameters simultaneously. This proves valuable when boundary conditions are uncertain (Panday et al. 2025).

Data assimilation frameworks systematically combine model predictions with observations. Ensemble Kalman Filters propagate uncertainty while updating predictions based on observations, enabling real-time updating critical for operational forecasting. Residual modelling uses data-driven methods to correct systematic biases in process-based predictions, maintaining interpretability while empirically accommodating simplified processes. Hybrid approaches offer improved accuracy, enhanced reliability through physical constraints, and better uncertainty characterisation. However, implementation requires dual expertise, and validation becomes complex when model types interact.

3.3.3 Empirical Models

Empirical models occupy a simpler position within the modelling spectrum, using established statistical relationships and classical formulations to predict contamination patterns. These approaches suit situations with limited real-time data where historical relationships provide adequate predictive power. Traditional statistical methods form the core of empirical approaches. Multiple linear regression remains popular for its simplicity, treating contamination as a linear combination of predictors such as flow and temperature. Generalized linear models handle non-normal distributions, while generalized additive models capture non-linear effects while maintaining interpretability.

Time series models explicitly account for temporal dependencies. ARIMA models effectively forecast short-term patterns by decomposing series into trend, seasonal, and random components. Vector autoregression handles multiple interrelated contaminants, while state-space models combine process knowledge with statistical updating. Classical empirical formulations, such as the Streeter-Phelps equation for dissolved oxygen sag curves, represent well-established relationships derived from fundamental observations. These provide rapid, interpretable predictions suitable for screening-level assessments and situations where data or computational resources are limited. Empirical models offer accessibility, computational efficiency, and interpretability. However, assumptions of linear or monotonic relationships may inadequately capture complex dynamics, and applicability is restricted to conditions similar to those from which relationships were derived.

3.3.4 Process-Based Models

Process-based models simulate contaminant transport using mathematical representations of physical, chemical, and biological processes. Their primary strength lies in providing mechanistic insights that enable scenario analysis beyond the range of observational data, making them particularly suitable for long-term planning in complex systems (Sprocati and Rolle 2021; Liu et al. 2026). Hydrodynamic models form the foundation by solving flow equations. One-dimensional models like HEC-RAS prove effective for emergency response. Two-dimensional models address significant lateral gradients, while three-dimensional models resolve vertical stratification, which is critical in lakes and estuaries.

Water quality models are built on hydrodynamic foundations. QUAL2K incorporates multiple constituents, including dissolved oxygen, nutrients, and pathogens. WASP provides flexible simulation through a modular structure, with its TOXI module addressing bioaccumulation. CE-QUAL-W2 excels in stratified systems. Groundwater models like MODFLOW with MT3D address subsurface transport, while PHREEQC incorporates geochemical speciation. Process-based models enable defensible extrapolation, support site-specific decisions, and address future scenarios, including those related to climate change. However, data requirements are extensive, computational demands limit real-time applications, and required expertise may limit accessibility. Despite limitations, they remain indispensable when a mechanistic understanding is crucial for long-term planning and management interventions.

3.4 Model Development and Implementation

The successful implementation of contaminant forecasting models requires systematic attention to data management, parameter estimation, and uncertainty characterization. Table 3.1 provides a comparison of contamination forecasting modelling approaches, moving from theoretical frameworks to operational systems, that demand practical strategies for handling real-world data limitations, computational constraints, and stakeholder requirements. This section addresses the critical steps in model development, from initial data collection through validation and uncertainty assessment, guiding practitioners implementing contamination prediction systems.

3.4.1 Data Requirements and Preprocessing

The foundation of predictive contamination models lies in comprehensive, quality-assured input data spanning hydrological variables, water quality parameters, meteorological conditions, and source characteristics. Data requirements vary by modelling approach: process-based models require detailed physical parameters, while data-driven approaches require extensive historical records. Early understanding of these requirements enables efficient collection planning and identifies critical gaps limiting model applicability.

Hydrological Data: Water movement fundamentally controls pollutant transport, making hydrological data essential. Stream discharge measurements at multiple locations provide boundary conditions and calibration data. Process-based models require cross-sectional geometry, bed slopes, and roughness coefficients to compute velocities and dispersion. Stage-discharge relationships convert water levels to flow rates, though changes in channel morphology can alter them. Groundwater systems need

Table 3.1 Comparison of contamination forecasting modelling approaches

Evaluation parameter	Process-based models	Data-driven models	Hybrid frameworks	Performance benchmark	Implementation scale	Resilience support	Primary data source	Model robustness	Tech maturity
Technical foundation									
Core logic	Physical/ chemical laws	Statistical patterns	Physics-guided	Deterministic	System-wide	High	Lab/field	High	Established
Math basis	ADE and mass balance	Non-linear algorithms	Physics-informed NN	Theoretical	Foundational	Low	Coefficients	Moderate	Established
Interpretability	High: mechanistic	Low: black box	Medium insight	Explicit	High expertise	Medium	Expert knowledge	High	High
Architecture									
Example models	QUAL2K, WASP, SWAT	Random Forest, LSTM	PINNs	Standardized	Modular	Variable	Literature	High	Mature
Emerging tech	3D hydrodynamics	Graph neural Nets	Digital twins	Cutting-edge	Integrated	High	IoT streams	Moderate	Cutting-edge
Extrapolation	Good: beyond range	Poor: training range	Moderate	Defensible	Scenario-led	Low	Archives	Low	Developing
Operations									
Primary Input	Physical parameters	Historical datasets	Dual source data	Parameter-heavy	Intensive	Moderate	In-situ data	Moderate	High
Calibration	Parameter optimization	Hyperparameter tuning	Dual calibration	Optimization	Site-specific	Low	Telemetry	Low	Standard
Comp load	High: hours to days	Fast: real-time	Moderate	Resource-bound	HPC required	High	Cloud DBs	High	High
Validation									

(continued)

Table 3.1 (continued)

Evaluation parameter	Process-based models	Data-driven models	Hybrid frameworks	Performance benchmark	Implementation scale	Resilience support	Primary data source	Model robustness	Tech maturity
Uncertainty	Monte Carlo/ Morris	Ensemble methods	Bayesian/dual	Probabilistic	Multi-source	Medium	Error dist	Moderate	Mature
Primary error	Structural/ parameter	Data quality gaps	Interaction complexity	Diagnostic	Evaluative	High	Metadata	High	Emerging
Metrics	NSE, RMSE, Bias	Pattern recognition	Reliability/ accuracy	Statistical	Regulatory	Adaptive	Validation sets	High	Standard
Strategy									
Time horizon	Long-term: scenarios	Short-term: forecasts	Dual: multiscale	Scenario-based	Strategic	High	Climatology	Moderate	Mature
Best use case	Remediation design	Emergency response	Operational DSS	Proactive	Management	High	Event logs	High	High
Carbon link	Embodied carbon	Energy optimization	Life cycle nexus	Net-zero goal	Holistic	High	Carbon logs	High	Emerging
Ecology									
NbS integration	Low: hard to model	High: observed data	Medium: hybridized	Eco-centric	Restorative	Very High	Biota health	Moderate	Developing
Climate stress	Sensitivity analysis	Trend extrapolation	Dynamic adaptation	Stress-tested	Resilient	High	Temp/ precip	High	High
Governance									
Decision role	Regulatory policy	Tactical response	Risk-based	Goal-oriented	Institutional	High	Policy docs	Moderate	High

(continued)

Table 3.1 (continued)

Evaluation parameter	Process-based models	Data-driven models	Hybrid frameworks	Performance benchmark	Implementation scale	Resilience support	Primary data source	Model robustness	Tech maturity
Investment	High: initial/Op	Moderate: initial	High: specialized	Value-driven	Institutional	High	CAPEX logs	High	Mature
Social equity	Low: tech-focused	Moderate: bias risks	High: participatory	Fairness-based	Community	High	Surveys	High	Emerging

hydraulic head measurements, aquifer properties, and pumping rates. Temporal resolution must match contamination dynamics—hourly data for urban runoff events, daily or weekly for groundwater applications.

Water Quality Monitoring: Concentration data drives model calibration and validation. Basic parameters (temperature, pH, dissolved oxygen, and conductivity) provide general insights into water quality and may serve as surrogates for unmeasured constituents (Kumar et al. 2023; Agarwal et al. 2025). Nutrients are critical for eutrophication modelling, while heavy metals, pesticides, and emerging contaminants require specialized analytics with appropriate detection limits. Spatial distribution should capture sources, tributaries, and gradients; temporal frequency must resolve seasonal patterns and episodic events. Continuous sensors increasingly supplement grab sampling, capturing short-term variability.

Meteorological Data: Weather influences contaminant fate through multiple pathways. Temperature affects reaction rates; solar radiation drives photodegradation and primary production; precipitation mobilizes land-based contaminants and alters dilution; wind affects mixing and volatilization. Evapotranspiration data close water balances for flow estimation. Station proximity and representativeness significantly impact accuracy—complex terrain or large water bodies may require multiple stations.

Source Characterization: Quantifying inputs requires distinguishing point and non-point sources. Point sources (wastewater plants, industries) are subject to permit-required monitoring, though actual loadings may differ from permitted values. Combined sewer overflows and stormwater present challenges due to intermittency and variability. Non-point sources require indirect estimation using land use data, fertilizer applications, livestock densities, impervious surface mapping, and traffic patterns. Chemical fingerprinting or isotopic analysis helps identify diffuse inputs.

Data Preprocessing and Quality Control: Raw data must be transformed into model-ready formats while addressing quality issues. Missing data treatment ranges from linear interpolation for short gaps to regression imputation for extended periods. Outlier detection using statistical methods identifies erroneous values, though extreme observations may represent genuine contamination events that require expert judgment. QA/QC procedures ensure reliability through laboratory blanks, duplicates, spike recoveries, field replicates, and equipment blanks. Data validation includes range tests, consistency assessments, and mass balance evaluations.

Data transformation improves model performance; log-transformation normalizes skewed distributions; and standardization enables combining variables across different scales. Temporal aggregation from high-frequency sensors requires appropriate averaging methods (flow-weighted for concentrations). Spatial interpolation generates continuous fields from point measurements using methods from simple inverse distance weighting to sophisticated geostatistical approaches. Metadata documentation and version control ensure reproducibility and support collaborative development.

3.4.2 Model Calibration and Validation

Model calibration adjusts parameters to minimize discrepancies between simulated and observed values, transforming generic model structures into site-specific predictive tools. The calibration process requires careful consideration of parameter identifiability, objective function selection, and optimization methods. Effective calibration strategies balance model complexity with available data, avoiding overparameterization that reduces predictive reliability despite improving calibration fits. Parameter selection for calibration begins with a sensitivity analysis that identifies influential parameters warranting adjustment. Local sensitivity analysis examines parameter impacts near nominal values, while global methods explore the entire parameter space. The Morris method efficiently screens many parameters (Nabi et al. 2021), while Sobol indices quantify parameter interactions (Langstaff et al. 2022). Parameters exhibiting minimal sensitivity can be fixed at literature values, reducing calibration complexity. Highly sensitive parameters require careful calibration but may be difficult to identify if their effects are similar, necessitating additional data or model simplification.

Objective function selection critically influences calibration outcomes by defining the measure of model-data agreement. The Nash–Sutcliffe efficiency (NSE), which compares model predictions to the observed mean, is a widely used normalized metric in hydrological applications (Melsen et al. 2025). Root mean square error (RMSE) emphasizes large errors, which is appropriate when extreme concentrations are management priorities. Mean absolute error (MAE) treats all errors equally, providing robustness to outliers. Percent bias indicates systematic over- or underprediction, which is important for mass balance assessment. Multi-objective calibration accounts for trade-offs among different performance measures, yielding Pareto-optimal parameter sets that inform uncertainty bounds (Harkare et al. 2024).

Optimization algorithms search the parameter space for values minimizing objective functions. Gradient-based methods like Levenberg–Marquardt converge rapidly for smooth objective functions but may become trapped in local optima. Global optimization methods, such as genetic algorithms and particle swarm optimization, explore the parameter space more thoroughly but require many model evaluations. Bayesian methods like DREAM (Differential Evolution Adaptive Metropolis) simultaneously optimize parameters and quantify uncertainty through posterior parameter distributions. The choice of optimization method depends on the computational demands of the model, the parameter dimensionality, and the need for uncertainty quantification.

Calibration strategies should reflect intended model applications and data characteristics. Split-sample calibration reserves data subsets for validation, though limited data may necessitate cross-validation approaches. Differential split-sample tests evaluate model transferability by calibrating to different hydrological conditions. Multi-site calibration improves parameter identifiability and spatial transferability by simultaneously fitting multiple locations. Sequential calibration first adjusts flow

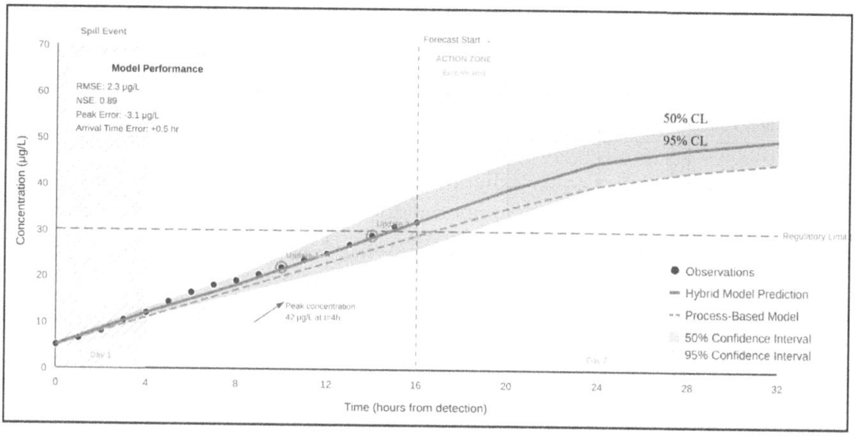

Fig. 3.3 Time series prediction demonstrating contamination forecasting model performance for a chemical spill event

parameters before water quality parameters, acknowledging the hierarchical dependency. Regularization techniques, including Tikhonov regularization and parameter penalties, prevent overfitting by favouring simpler models with parameter values close to prior estimates. The practical application of model calibration and validation procedures is illustrated through a contamination forecasting example involving tetrachloroethylene (PCE) transport following an industrial spill (Fig. 3.3).

The hybrid model, calibrated using 16 h of observations, demonstrates excellent performance with a Nash–Sutcliffe efficiency of 0.89 and RMSE of 2.3 μg/L. The validation reveals critical insights into model behaviour: peak concentration predictions show a slight negative bias of −3.1 μg/L, while arrival time estimates exhibit a positive error of +0.5 h, both within acceptable ranges for operational decision-making. Real-time data assimilation at 6- and 10-h post-detection significantly improves forecast accuracy, reducing prediction uncertainty by approximately 25% as indicated by the narrowing confidence intervals around update points. The comparison with a traditional process-based model highlights the hybrid approach's superior performance, particularly in capturing the contamination decay phase. Notably, the expanding uncertainty bounds during the forecast period (16–32 h) appropriately reflect decreasing confidence in longer-term predictions, providing water managers with realistic assessments of forecast reliability. This example demonstrates how systematic validation not only quantifies model performance but also identifies opportunities for improvement through targeted data collection and model refinement.

Model validation assesses predictive capability using independent data not used in calibration. Temporal validation applies calibrated models to different time periods, testing robustness to changing conditions. Spatial validation evaluates transferability to different locations within the study area. Proxy validation uses related variables to assess model consistency, such as comparing simulated temperature with dissolved oxygen, given their coupling. Event-based validation focuses on specific contamination episodes critical for management decisions. Statistical validation tests compare observed and simulated distributions rather than point-by-point correspondence, appropriate for stochastic predictions.

Performance metrics for validation extend beyond calibration statistics to include management-relevant measures. Reliability indicates the frequency of correct predictions for threshold exceedances important for regulatory compliance. Resilience measures recovery from poor predictions, relevant for adaptive management systems. Resolution quantifies the model's ability to distinguish between contamination levels. Discrimination assesses the categorical prediction accuracy of, for example, beach closure recommendations. Economic metrics translate prediction errors into monetary impacts, supporting cost–benefit analysis of model improvements.

3.4.3 Uncertainty Analysis

Uncertainty permeates all aspects of contaminant forecasting, from measurement errors through model structural limitations to future scenarios. Comprehensive uncertainty analysis quantifies prediction reliability, identifies dominant uncertainty sources, and communicates confidence bounds essential for risk-based decision-making. Input Uncertainty: Measurement errors, spatial interpolation, and temporal gaps all contribute to input uncertainty. Analytical precision, sampling representativeness, and transcription errors contribute to measurement uncertainty. Spatial uncertainty arises when point measurements represent larger areas or interpolation fills unmonitored regions. Temporal uncertainty results from assuming constant values between measurements. Monte Carlo simulation assesses propagation by generating input ensembles sampling from error distributions.

Parameter Uncertainty: Model coefficients remain uncertain even after calibration, with parameters varying naturally (like microbial degradation rates) or remaining unknown for emerging contaminants. Calibration yields estimates but not unique solutions—different combinations produce similar fits. The equifinality concept acknowledges that multiple valid parameter sets are possible given data limitations. Bayesian calibration quantifies uncertainty via posterior distributions, whereas frequentist approaches use bootstrap confidence intervals. Parameter correlation must be considered, as compensating errors may artificially reduce apparent uncertainty. Structural Uncertainty: Model simplifications, numerical approximations, and excluded processes create structural uncertainty. All models make assumptions about spatial dimensionality, temporal resolution, and process complexity. Numerical solutions introduce discretization errors; missing processes create systematic biases. Multi-model ensembles address this by combining predictions from different formulations. Bayesian model averaging weights models by posterior probability, with spread among models measuring structural uncertainty, though shared biases may cause underestimation (Panday et al. 2024).

Future conditions that affect predictions include climate change (precipitation patterns, temperature regimes, and extreme events), land-use changes, regulatory changes, and demographic shifts. Scenario analysis explores plausible futures through different development storylines. Robust decision-making identifies strategies performing adequately across multiple scenarios rather than optimally for single futures. Uncertainty Propagation and Communication: Monte Carlo simulation remains the most general propagation approach, repeatedly running models with sampled inputs. Latin Hypercube sampling improves efficiency; first-order methods provide rapid linear approximations; polynomial chaos expansion enables efficient propagation for smooth models. Computational costs often exceed deterministic modelling by orders of magnitude, necessitating surrogate modelling or parallel computing.

Communication challenges involve conveying probabilistic information to diverse audiences. Confidence intervals provide simple bounds; exceedance curves show threshold probabilities; ensemble plots display multiple realizations. Color-coded maps reveal spatial patterns, while interactive visualizations enable exploration of uncertainty. Effective communication requires understanding audience needs and iteratively refining presentations. Implementation balances theoretical completeness with practical constraints, including data limitations, computational resources, time constraints, and regulatory requirements. Investment in robust uncertainty characterization yields increased stakeholder confidence, better risk management, and improved environmental outcomes.

3.5 Integration with Decision Support Systems

Transforming contaminant forecasting models from technical tools into operational decision-support systems requires careful integration of predictive capabilities with management workflows, stakeholder needs, and institutional constraints. Effective decision support systems (DSS) bridge the gap between complex model outputs and actionable management decisions, providing timely, relevant, and understandable information to diverse users, ranging from water treatment operators to policymakers.

3.5.1 Early Warning Systems and Risk Assessment

Early warning systems represent the most critical application of contamination forecasting, providing advance notice of water quality threats to enable protective actions. Effective systems integrate real-time monitoring networks, predictive models, alert generation algorithms, and communication protocols. The system must balance sensitivity to detect genuine threats against specificity to minimize false alarms that erode user confidence. Modern systems employ tiered alert levels (Fig. 3.4) reflecting

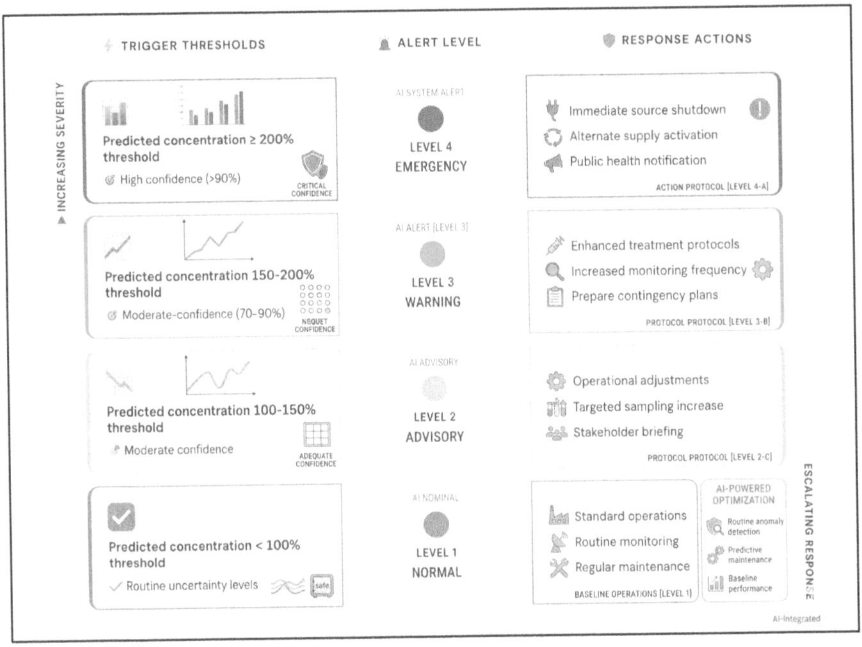

Fig. 3.4 Tiered early warning system response framework showing escalating alert levels based on predicted contamination thresholds and forecast confidence, with corresponding trigger criteria and management responses

both contamination severity and prediction uncertainty, enabling graduated responses proportional to risk levels.

The technical infrastructure requires robust data acquisition and processing capabilities. Automated sensors transmit measurements through telemetry networks to centralized databases, with redundant communication pathways ensuring reliability during emergencies. Data validation algorithms identify sensor malfunctions that could compromise performance. The forecasting engine runs continuously, comparing predictions against threshold values derived from regulatory standards and health-based criteria. When exceedances are predicted, the system evaluates uncertainty bounds to determine alert confidence levels.

Alert dissemination must reach appropriate recipients through reliable channels with minimal delay. Multi-modal communication via email, SMS, automated calls, and mobile applications ensures message delivery even when individual channels fail. Alert messages are tailored to recipient roles: operators receive detailed technical information, while officials receive summarized risk assessments. Escalation protocols automatically notify higher management for severe events or when primary recipients don't acknowledge alerts within specified timeframes.

Contamination forecasts also provide essential inputs for quantitative risk assessment, enabling proactive evaluation of potential health and ecological impacts. Integration requires translating model outputs into exposure estimates that account for multiple pathways, including drinking water consumption, recreational contact, and food chain transfer. Probabilistic forecasts support risk characterization by providing exposure distributions rather than point estimates. Risk-based decision support employs forecasts to optimize management interventions, with multi-criteria frameworks incorporating predicted risk reductions, implementation costs, and technical feasibility to rank alternative responses.

3.5.2 Operational Decision Support Tools

Water treatment plant operators require tools that translate contamination forecasts into operational guidance. Treatment optimization modules use predicted influent quality to recommend process adjustments such as coagulant doses, oxidant addition rates, or filter run times. Machine learning models trained on historical plant performance predict finished water quality under different operational scenarios. Cost-optimization algorithms balance treatment effectiveness with chemical and energy costs, particularly during extended contamination periods. Reservoir and river system management benefits through improved operational planning. Selective withdrawal strategies in stratified reservoirs use depth-specific quality predictions to minimize the release of contaminated water.

Flow augmentation decisions balance dilution benefits against water conservation needs, with forecasts enabling precise timing determinations. Emergency drawdown planning uses transport predictions to maximize contaminant flushing while minimizing downstream impacts. Source water protection programs leverage forecasting to prioritize monitoring and intervention activities. Predicted vulnerability windows guide targeted sampling campaigns, maximizing detection probability while minimizing costs. Watershed patrol scheduling focuses on high-risk periods identified through forecasting and improving violation detection. Best management practice implementation targets critical source areas based on predicted contaminant generation and transport.

3.5.3 Stakeholder Communication and Policy Support

Effective systems must communicate complex forecasting information to diverse stakeholders with varying technical backgrounds. Web-based dashboards provide customizable interfaces that enable users to access relevant information at appropriate levels of detail. Map-based visualizations show predicted contamination plumes with uncertainty bounds, while time-series plots compare forecasts with historical ranges and regulatory thresholds. Interactive features allow users to explore scenarios and understand the sensitivity of predictions. Public communication requires a balance between transparency and the avoidance of unnecessary alarm. Simplified indicator systems using color codes convey general threat levels without technical complexity.

Mobile applications extend accessibility, enabling field staff to access forecasts and submit observations from any location, even when offline, while offline capabilities ensure functionality during network outages.

Contamination forecasting systems increasingly support regulatory compliance and policy development. Automated reporting modules generate required documentation for discharge permits, demonstrating proactive management. Forecasting systems provide evidence for enforcement actions by reconstructing historical events and identifying likely sources. Total maximum daily load development uses models to evaluate allocation scenarios, while climate change adaptation planning employs forecasts to evaluate infrastructure vulnerability under future conditions. The integration of contamination forecasting with decision support systems transforms reactive water quality management into proactive protection strategies. Success requires technical capability, institutional commitment, stakeholder engagement, and sustained resources for system maintenance.

3.6 Future Directions and Challenges

The field of contaminant forecasting stands at a transformative juncture where technological advances, increasing environmental pressures, and evolving societal expectations converge to create both unprecedented opportunities and complex challenges. Emerging technologies, including artificial intelligence, Internet of Things sensors, and digital twin systems, promise revolutionary improvements in predictive capabilities, while climate change and novel contamination sources demand adaptive approaches beyond current methodologies.

3.6.1 Artificial Intelligence and Big Data Integration

The exponential growth in environmental data from satellite observations, sensor networks, and citizen science creates opportunities for AI-driven contamination forecasting that surpasses current capabilities. Deep learning architectures, particularly transformer models that have revolutionized natural language processing, show promise for capturing complex spatiotemporal dependencies in water quality data. These models can process multiple data streams simultaneously, identifying subtle patterns invisible to traditional analysis. Graph neural networks naturally represent river networks and distribution systems, propagating information along flow paths while maintaining physical connectivity constraints. Reinforcement learning approaches optimize monitoring and treatment strategies by learning from historical decisions and outcomes, potentially discovering non-intuitive solutions that outperform conventional management.

The challenge lies not in data scarcity but in effectively leveraging heterogeneous, noisy, and incomplete datasets. Data fusion techniques must reconcile differences in spatial and temporal resolution, measurement uncertainties, and semantic inconsistencies across sources. Automated quality control using anomaly detection algorithms helps identify erroneous data before model training, though distinguishing genuine extreme events from errors remains problematic. The "black box" nature of complex AI models raises concerns about interpretability and trust, particularly for high-stakes decisions affecting public health.

Edge computing enables real-time forecasting at monitoring locations without relying on centralized processing. Smart sensors with embedded AI capabilities can detect contamination events, trigger enhanced monitoring, and generate local predictions autonomously. Federated learning allows multiple organizations to collaboratively train models without sharing sensitive data, addressing privacy concerns while leveraging collective information. These distributed approaches enhance system resilience against cyber-attacks and communication failures during emergencies.

3.6.2 Climate Change and Emerging Contaminants

Climate change fundamentally alters the boundary conditions underlying current forecasting models. Non-stationarity in hydrological regimes invalidates historical statistics used for model calibration and uncertainty quantification. Extreme events outside historical experience challenge model extrapolation capabilities. Temperature increases affect reaction kinetics, stratification patterns, and biological processes, cascading through water quality dynamics. Sea level rise alters salinity intrusion and contaminant transport in coastal systems, requiring coupled modelling of previously independent processes.

Compound events, in which multiple stressors coincide, create contamination scenarios that exceed the sum of their individual impacts. Heat waves combined with drought concentrate pollutants while accelerating algal growth and toxin production. Wildfires followed by intense rainfall mobilize ash-bound contaminants in amounts overwhelming treatment systems. Current models, typically calibrated for single-stressor responses, may severely underestimate risks from compound events, requiring new mathematical frameworks that capture stressor interactions and threshold effects.

The accelerating pace of chemical innovation continuously introduces novel contaminants lacking toxicological data and environmental fate information. Pharmaceutical metabolites, microplastics, engineered nanomaterials, and synthetic biology products exhibit unique transport behaviours not captured by conventional models. Predictive modelling increasingly relies on computational chemistry and quantitative structure–activity relationships to estimate properties before environmental release. Physics-informed neural networks trained on well-studied compounds extrapolate to novel chemicals with similar structures.

3.6.3 Digital Twins and Research Priorities

Digital twins, continuously updated virtual replicas of water systems, represent the future of integrated contamination forecasting and management. These cyber-physical systems combine high-fidelity simulation models with real-time data streams to mirror actual system states and predict future evolution. Unlike traditional models that run periodically, digital twins operate continuously, assimilating observations to maintain synchronization with the physical systems they model. Contamination events detected in the physical system are immediately reflected in the digital twin, which forecasts downstream impacts and evaluates responses.

Digital twins require tightly integrated data systems, simulation models, and control infrastructure, supported by cloud computing resources capable of performing ensemble simulations and uncertainty quantification. Critical research gaps constrain the advancement of contamination forecasting capabilities. Fundamental understanding of processes remains incomplete for many biogeochemical interactions. Scaling relationships linking laboratory measurements to field-scale parameters need systematic investigation. Uncertainty quantification methods must evolve beyond parametric approaches to address deep uncertainties in model structures and future scenarios.

Institutional challenges may prove more limiting than technical constraints. Data-sharing agreements require legal frameworks that balance transparency with security concerns. Workforce development must prepare practitioners to implement increasingly sophisticated forecasting tools. Public trust in AI-driven predictions requires transparent development processes and demonstrated reliability. The convergence of technological capabilities and environmental pressures positions contamination forecasting for transformative advances, requiring coordinated efforts among researchers, practitioners, and policymakers.

3.7 Conclusions

Predictive modelling for contaminant forecasting has emerged as a central component of contemporary water quality management, enabling the anticipation of contamination risks and informing timely intervention. As discussed in this chapter, the effectiveness of such forecasting systems lies in integrating mechanistic understanding with data-driven analytics, supported by expanding monitoring networks and computational capabilities. Process-based models remain indispensable for physical interpretation and scenario testing, while data-driven and hybrid frameworks offer practical advantages for operational prediction in complex, data-rich environments. Nevertheless, the value of predictive modelling depends not merely on technical sophistication but on appropriate model selection, transparent uncertainty characterization, and the ability to translate modelling outputs into decision-relevant information for managers and stakeholders.

Equally important is the recognition that contamination forecasting functions within a broader institutional and environmental context. Successful implementations consistently demonstrate that robust monitoring infrastructure, iterative model refinement, and long-term institutional commitment are as critical as modelling techniques themselves. When integrated with governance frameworks and proactive water management strategies, forecasting tools enable a shift from reactive remediation toward preventive protection of water resources. Looking ahead, advances in artificial intelligence, real-time sensing, and digital water systems will further strengthen predictive capabilities, yet their effectiveness will ultimately depend on sustained collaboration among scientists, policymakers, and water managers to ensure that modelling insights translate into resilient and sustainable water quality management.

References

Agarwal V, Kumar A, Gomes RL, Marsh S (2020) Monitoring of ground movement and groundwater changes in London using InSAR and GRACE. Appl Sci 10(23):8599

Agarwal V, Kumar A, Gee D, Grebby S, Gomes RL, Marsh S (2021) Comparative study of groundwater-induced subsidence for London and Delhi using PSInSAR. Remote Sens 13(23):4741

Agarwal V, Kumar M, Dogra K, Mejia-Avendaño S (2025) Isomers of perfluorooctanesulfonate exhibit preferential infiltration and contrasting ionic associations between surface water and groundwater. J Hazard Mater 494:138445

Amrutha, Jeppu G, Girish CR, Prabhu B, Mayer K (2023) Multi-component adsorption isotherms: review and modeling studies. Environ Process 10(2):38

Banerjee A, Creedon L, Jones N, Gill L, Gharbia S (2023) Dynamic groundwater contamination vulnerability assessment techniques: a systematic review. Hydrology 10(9):182

Carrera J, Saaltink MW, Soler-Sagarra J, Wang J, Valhondo C (2022) Reactive transport: a review of basic concepts with emphasis on biochemical processes. Energies 15(3):925

Ciampi P, Esposito C, Cassiani G, Deidda GP, Flores-Orozco A, Rizzetto P et al (2022) Contamination presence and dynamics at a polluted site: spatial analysis of integrated data and joint conceptual modeling approach. J Contam Hydrol 248:104026

Haghighat M, Nikoo MR, Talebbeydokhti N, Hassani MR (2026) Reservoir water quality planning based on a robust multi-criteria decision-making approach under deep uncertainty. J Environ Manage 397:128271

Harkare V, Mangrulkar R, Thorat O, Jain SR (2024) Evolutionary approaches for multi-objective optimization and pareto-optimal solution selection in data analytics. In: Applied multi-objective optimization. Springer Nature Singapore, Singapore, pp 67–94

Kumar A, Nehdi ML (2026) Data-driven approaches to groundwater modelling: methods, applications, and challenges. Hydrol Insights 1–10

Kumar M, Panday DP, Bhagat C, Herbha N, Agarwal V (2023) Demystifying the decadal shift in the extent of groundwater in the coastal aquifers of Gujarat, India: a case of reduced extent but increased magnitude of seawater intrusion. Sci Tot Environ 898:165451

Langstaff J, Glen G, Holder C, Graham S, Isaacs K (2022) A sensitivity analysis of a human exposure model using the Sobol method. Stoch Env Res Risk Assess 36(11):3945–3960

Lin F, Ren H, Qin J, Wang M, Shi M, Li Y et al (2024) Analysis of pollutant dispersion patterns in rivers under different rainfall based on an integrated water-land model. J Environ Manage 354:120314

Linnik P, Osadchyi V, Osadcha N, Linnik R (2023) Redox potential as an important characteristic of the chemical and biological state of surface waters. Chem Ecol 39(6):640–672

Liu Y, He Y, Han B, Liu H, Tao S, Liu W (2023) Sewage discharge and organic matter affect the partitioning behaviors of different polycyclic aromatic hydrocarbons in a river surface sediment-pore water system. J Hazard Mater 446:130757

Liu S, Li Y, Zhan C, Liu H, Wang Y, Fang L (2026) A review of pollutant leaching from tailings: numerical modeling, transport mechanisms, and pollution control. Environ Rev (ja)

Masinga P, Simbanegavi TT, Makuvara Z, Marumure J, Chaukura N, Gwenzi W (2024) Emerging organic contaminants in the soil–plant-receptor continuum: transport, fate, health risks, and removal mechanisms. Environ Monit Assess 196(4):367

Mathur K, Sharma P, Gaur N, Kumar S (2025) Enhancing water quality management: the role of predictive modeling and IoT in monitoring, analysis, and intervention. In Computational automation for water security. Elsevier, pp 43–68

Melsen LA, Puy A, Torfs PJ, Saltelli A (2025) The rise of the Nash-Sutcliffe efficiency in hydrology. Hydrol Sci J 1–12

Miller T, Durlik I, Kostecka E, Kozlovska P, Łobodzińska A, Sokołowska S, Nowy A (2025) Integrating artificial intelligence agents with the internet of things for enhanced environmental monitoring: applications in water quality and climate data. Electronics 14(4):696

Mishra V, Malik K, Agarwal V, Mishra PK, Jain K (2024) Impact assessment of unsustainable airport development in the Himalayas using remote sensing: a case study of Pakyong Airport, Sikkim, India. Quat Sci Adv 13:100144

Nabi S, Ahanger MA, Dar AQ (2021) Investigating the potential of Morris algorithm for improving the computational constraints of global sensitivity analysis. Environ Sci Pollut Res 28(43):60900–60912

Panday DP, Kumar M (2024) Decoding uranium variation over the Indian Peninsula through leveraging standardized precipitation evapotranspiration indices and groundwater level fluctuations. Groundw Sustain Dev 26:101241. https://doi.org/10.1016/J.GSD.2024.101241

Panday DP, Kumar M, Agarwal V, Torres-Martínez JA, Mahlknecht J (2024) Corroboration of arsenic variation over the Indian Peninsula through standardized precipitation evapotranspiration indices and groundwater level fluctuations: water quantity indicators for water quality prediction. Sci Total Environ 954:176339. https://doi.org/10.1016/j.scitotenv.2024.176339

Panday DP, Kumari A, Kumar M (2025) Alkalinity-salinity-sustainability: decadal groundwater trends and its impact on agricultural water quality in the Indian Peninsula. Sci Total Environ 978:179459. https://doi.org/10.1016/j.scitotenv.2025.179459

Qin Z, Agarwal V, Gee D, Marsh S, Grebby S, Chen Y, Meng N (2021) Study of ground movement in a mining area with geological faults using FDM analysis and a stacking InSAR method. Front Environ Sci 9:787053

Saadat AM, Khodambashi Emami S, Hamidifar H (2024) A review on storage process models for improving water quality modeling in rivers. Hydrology 11(11):187

Sanskrityayn A, Suk H, Chen JS, Park E (2021) Generalized analytical solutions of the advection-dispersion equation with variable flow and transport coefficients. Sustainability 13(14):7796

Seyedpour SM, Thom A, Ricken T (2023) Simulation of contaminant transport through the vadose zone: a continuum mechanical approach within the framework of the extended theory of porous media (etpm). Water 15(2):343

Sigmund G, Arp HPH, Aumeier BM, Bucheli TD, Chefetz B, Chen W et al (2022) Sorption and mobility of charged organic compounds: how to confront and overcome limitations in their assessment. Environ Sci Technol 56(8):4702–4710

Singh BVR, Agarwal V, Sanwal V (2024) Climatic shifts and vegetation response in Western India: a four-decade retrospective through GIS and multi-variable analysis. Oxf Open Clim Change

Sprocati R, Rolle M (2021) Integrating process-based reactive transport modeling and machine learning for electrokinetic remediation of contaminated groundwater. Water Resour Res 57(8):e2021WR029959

Talukdar P, Kumar B, Kulkarni VV (2023) A review of water quality models and monitoring methods for capabilities of pollutant source identification, classification, and transport simulation. Rev Environ Sci Biotechnol 22(3):653–677

van der Heijden T, Mendoza-Lugo MA, Palensky P, van de Giesen N, Abraham E (2025) Incorporating risk in operational water resources management: probabilistic forecasting, scenario generation, and optimal control. Water Resour Res 61(3):e2024WR037115

Zhang Y, Zhou D, Wei W, Frame JM, Sun H, Sun AY, Chen X (2021) Hierarchical fractional advection-dispersion equation (FADE) to quantify anomalous transport in river corridor over a broad spectrum of scales: theory and applications. Mathematics 9(7):790

Zhang T, Feng M, Chen K (2023) The quantitative response of pollutant spatial distribution and mixing dynamics to confluent conditions in an asymmetric open-channel confluence. J Hydrol 620:129479

Zhou T, Gani R, Sundmacher K (2021) Hybrid data-driven and mechanistic modeling approaches for multiscale material and process design. Engineering 7(9):1231–1238

Chapter 4
Enhancing the Water Resilience with Special Emphasis on Carbon Neutrality

Abstract Water and carbon systems are inextricably linked in sustainable development, presenting the dual challenge of ensuring water security while decarbonizing economies to mitigate climate change. The water sector accounts for approximately 4% of global electricity consumption, rising to much higher levels in water-scarce regions that rely on energy-intensive solutions such as desalination and deep groundwater extraction. Traditional water infrastructure, including dams, long-distance transfers, and intensive treatment facilities, often contributes to climate change that threatens water systems, creating a problematic feedback loop. The water-carbon nexus framework demonstrates how water management decisions have far-reaching carbon implications beyond direct energy consumption, including methane emissions from reservoirs and wastewater treatment processes. Nature-based solutions offer exemplary integrated approaches that simultaneously enhance flood resilience, improve water quality, support biodiversity, and sequester carbon through wetland restoration, watershed protection, and green infrastructure. The water-energy nexus reveals important synergies between renewable energy adoption and water conservation, as solar and wind power require far less water than thermal generation. Pathways to carbon–neutral water systems include emissions-reduction strategies, energy-efficiency improvements, renewable energy integration, and resource recovery from wastewater. Governance frameworks and policy mechanisms essential for achieving integrated water-carbon objectives encompass carbon pricing, regulatory standards, and stakeholder engagement. Practical guidance is provided for transitioning water infrastructure toward carbon neutrality while enhancing resilience against climate-induced hydrological changes.

Keywords Water-carbon nexus · Carbon neutrality · Water resilience · Nature-based solutions · Water-energy nexus · Climate change mitigation

M. Kumar et al., *Bridging Water Quality and Quantity in the Anthropocene*, SpringerBriefs in Water Science and Technology,
https://doi.org/10.1007/978-3-032-25655-3_4

4.1 Introduction: The Water-Carbon Nexus

Water and carbon, two elements fundamental to life on Earth, have become inextricably linked in the discourse of sustainable development. As the twenty-first century unfolds, humanity faces the dual challenge of ensuring water security for a growing global population while simultaneously decarbonizing economies to mitigate the worst effects of climate change. These challenges are not parallel concerns to be addressed in isolation; rather, they represent deeply interconnected systems whose fates are bound together in what scholars and practitioners increasingly refer to as the water-carbon nexus (Panday et al. 2025).

The concept of water resilience has emerged as a critical framework for understanding how communities, ecosystems, and economies can withstand and adapt to water-related stresses. At its core, water resilience encompasses the capacity of water systems to absorb disturbances, maintain essential functions during periods of stress, and transform in response to changing conditions. This includes resilience against floods and droughts, the ability to maintain water quality despite pollution pressures, and the institutional and infrastructural flexibility to respond to shifting demands (Jiang et al. 2022). However, traditional approaches to building water resilience have often overlooked a crucial dimension: the carbon intensity of the solutions themselves. Dams, desalination plants, long-distance water transfers, and energy-intensive treatment facilities may enhance water security in the short term, but their substantial carbon footprints contribute to the very climate change that threatens water systems in the first place (Rothausen and Conway 2011).

The relationship between water and carbon operates through multiple pathways that deserve careful examination. Figure 4.1 demonstrates the Water-Carbon Feedback Cycle. First, water systems are significant energy consumers and, consequently, substantial sources of greenhouse gas emissions (Yue et al. 2023). Globally, the water sector accounts for approximately 4% of electricity consumption, with this figure rising significantly in water-scarce regions that rely on energy-intensive sources such as desalination or deep groundwater extraction (Qin et al. 2024; Panday et al. 2026). The processes of abstracting water from rivers, lakes, and aquifers; treating it to potable standards; pumping it through distribution networks to homes and businesses; collecting wastewater through sewerage systems; and treating that wastewater before discharge or reuse collectively represent an enormous energy demand. In many countries, water utilities rank among the largest consumers of electricity within municipal operations, often accounting for 30–40% of a city's total energy use (Cardoso et al. 2023; Sushma et al. 2023).

Second, the climate change driven by carbon emissions fundamentally alters the hydrological cycle upon which water security depends. Rising temperatures increase evaporation rates, shift precipitation patterns, accelerate glacier melt, and intensify both droughts and floods. These changes do not affect all regions equally; some areas face increasing water scarcity while others contend with more frequent and severe flooding. Coastal aquifers face saltwater intrusion as sea levels rise, compromising freshwater supplies for millions of people. The timing of water availability changes

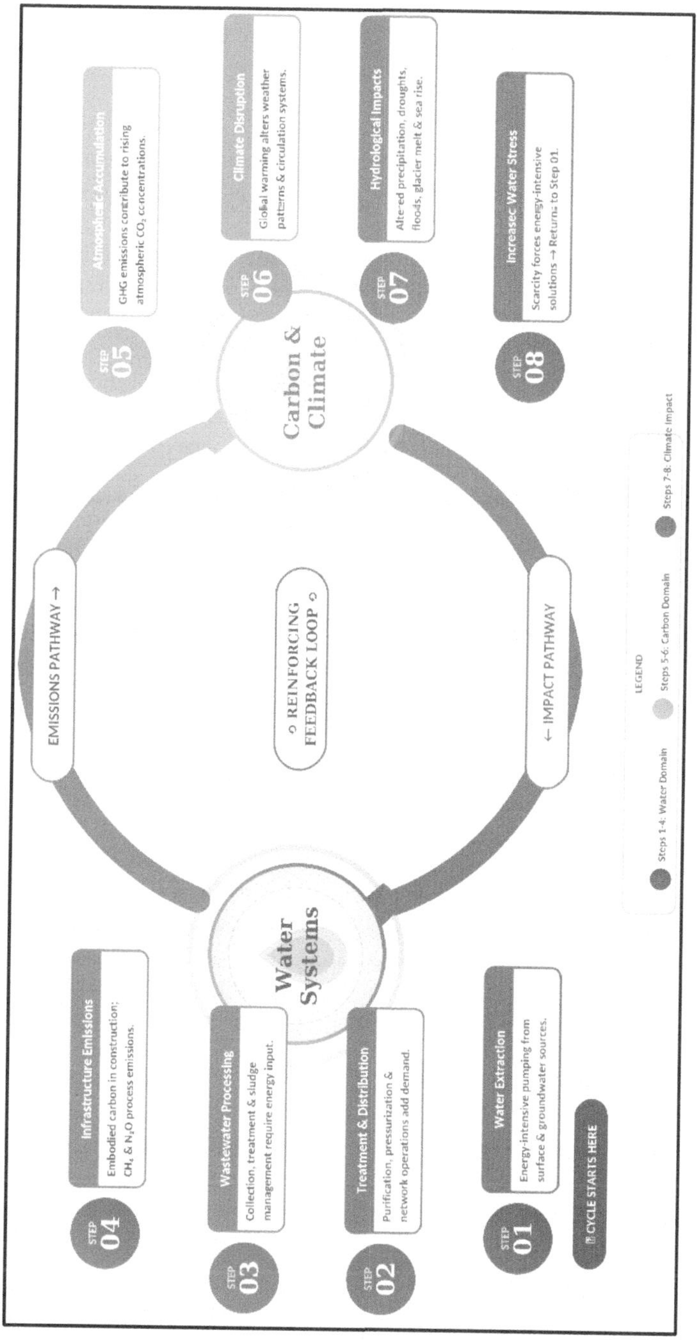

Fig. 4.1 The water-carbon feedback cycle. An eight-stage cyclic process illustrating the interconnected dynamics between water systems and carbon emissions, demonstrating how water management contributes to climate change, which in turn exacerbates water stress

as snowpacks diminish and spring melts occur earlier, disrupting agricultural systems that have evolved around predictable seasonal patterns. In this way, the carbon emissions from water infrastructure contribute to a feedback loop that undermines the very water security that infrastructure was designed to provide (Sinha et al. 2023; Huang et al. 2023).

Third, water bodies and water management practices themselves play roles in the global carbon cycle that extend beyond energy consumption. Reservoirs, particularly those in tropical regions, can be significant sources of methane emissions as organic matter decomposes in anaerobic conditions beneath the water surface. Wetlands, depending on their condition and management, act as either carbon sinks or sources (Li et al. 2024; Ye et al. 2022). Wastewater treatment processes release both carbon dioxide and methane, with the latter being a particularly potent greenhouse gas. Agricultural water use connects to carbon emissions through the production of nitrogen-based fertilizers, the operation of irrigation equipment, and the emissions from waterlogged rice paddies. Understanding these connections reveals that water management decisions have carbon implications that extend far beyond the electricity bill.

The imperative to address these interconnections has never been more urgent. The Paris Agreement established the goal of limiting global warming to well below 2 °C above pre-industrial levels, with efforts to limit it to 1.5° (Meinshausen et al. 2022; Chowdhury and Agarwal 2024). Achieving these targets requires rapid, far-reaching transitions across energy, land use, urban infrastructure, and industrial systems. Water infrastructure must be part of this transition. Simultaneously, the United Nations Sustainable Development Goals adopted in 2015 include explicit targets for clean water and sanitation, climate action, and affordable and clean energy. These goals recognize that progress in one area cannot come at the expense of another; sustainable development requires integrated solutions that advance multiple objectives simultaneously.

The concept of carbon neutrality has gained prominence as organizations, cities, and nations commit to achieving net-zero emissions. For the water sector, carbon neutrality means balancing any greenhouse gas emissions associated with water services through a combination of emissions reduction and carbon removal or offsetting (Lehtoranta et al. 2025). Some water utilities have already achieved or are approaching carbon neutrality, demonstrating that ambitious targets are technically feasible. However, the pathway to carbon–neutral water systems involves more than simply purchasing renewable energy or carbon credits. It requires fundamental rethinking of how water services are planned, designed, operated, and governed.

Integrated approaches to water resilience and carbon neutrality recognize that the most effective solutions often address both challenges simultaneously (Kyriakopoulos and Sebos 2023; Chen et al. 2022). Nature-based solutions exemplify this integration. Restoring wetlands can enhance flood resilience, improve water quality through natural filtration, provide habitat for biodiversity, and sequester carbon in soils and vegetation. Protecting forested watersheds safeguards water supplies while maintaining carbon stocks in trees and soil (Kumar et al. 2025a; Singh et al. 2025). Green infrastructure in cities, including rain gardens, permeable pavements, and urban forests, manages stormwater while reducing the urban heat island effect and sequestering carbon. These approaches contrast with traditional grey infrastructure, which typically requires significant energy for construction and operation while providing narrower, single-purpose benefits.

The water-energy nexus provides another lens through which to view these interconnections. Producing energy requires water for cooling thermal power plants, generating hydroelectricity, extracting and processing fossil fuels, and cultivating bioenergy crops. Conversely, providing water services requires energy as previously described. Climate change affects both sides of this equation: droughts reduce hydropower generation and require increased pumping of depleted groundwater, while heat waves strain water supplies and energy systems (Kumar et al. 2023). Transitioning to renewable energy sources can reduce the carbon footprint of water systems while, in many cases, also reducing water consumption associated with energy production. Solar photovoltaic and wind power require minimal water compared to thermal generation, creating synergies between decarbonization and water conservation.

This chapter explores these interconnections in depth, examining how water systems can be designed, managed, and governed to enhance resilience while achieving carbon neutrality. The following sections address the components of water resilience and the threats it faces; the sources of carbon emissions within water systems; the technologies, practices, and approaches that can reduce those emissions while maintaining or enhancing water services; the policy frameworks needed to support this transition; and the real-world examples that demonstrate what is possible. Throughout, the emphasis is on integration rather than trade-offs, seeking solutions that advance both water security and climate objectives.

The journey toward carbon–neutral water resilience is not merely a technical challenge but a societal one. It requires new forms of collaboration between water utilities, energy providers, urban planners, agricultural managers, and environmental agencies. It demands investment in infrastructure and innovation, supported by financing mechanisms that recognize the long-term value of sustainable solutions. It calls for governance arrangements that can bridge traditional sectoral boundaries and align incentives across different levels of government and between public and private actors. And it depends on public understanding and engagement, as the choices made about water and carbon will shape the world inherited by future generations.

The stakes of this endeavour are considerable. Water stress already affects over two billion people worldwide, a number projected to increase substantially as populations grow and climate change intensifies. The carbon budget remaining to achieve climate targets is rapidly depleting, requiring accelerated action across all sectors of the economy. Failure to address these challenges in an integrated manner risks either building water infrastructure that locks in high emissions for decades or pursuing decarbonization strategies that neglect water security. Neither outcome is acceptable. The alternative presented in this chapter is a pathway that recognizes water and carbon as interconnected challenges, requiring solutions that build resilience for both human communities and the planet's climate system.

As we proceed through the subsequent sections, readers will encounter both the complexities of these challenges and the innovative responses emerging from communities, utilities, researchers, and policymakers worldwide. The transition to carbon–neutral water resilience is already underway in many places, offering lessons and inspiration for broader application. This chapter aims to synthesize current knowledge, highlight best practices, identify remaining barriers, and chart a course toward a future where water security and climate stability reinforce rather than undermine each other. In doing so, it contributes to the larger project of sustainable development, recognizing that the choices made today about water and carbon will reverberate through generations to come.

4.2 Understanding Water Resilience

The concept of resilience has migrated from ecology and engineering into the mainstream of water management discourse over the past two decades. While traditional approaches to water management focused primarily on reliability, seeking to ensure a consistent supply through infrastructure designed for historical hydrological conditions, resilience thinking acknowledges that the future will not resemble the past. Climate change, population dynamics, land use transformations, and evolving societal expectations create conditions of deep uncertainty that demand new approaches. Water resilience, understood comprehensively, encompasses not merely the ability to withstand shocks but the capacity to adapt and transform in response to changing circumstances.

4.2.1 Components of a Resilient Water System

A resilient water system comprises multiple interconnected elements that together determine its capacity to function under stress. Physical infrastructure comprises the most visible components, including pipes, pumps, treatment plants, reservoirs, and distribution networks that deliver water services. However, infrastructure alone does not confer resilience. The design philosophy matters enormously: systems built with redundancy, modularity, and flexibility can continue functioning when individual components fail, while highly centralized and optimized systems may be efficient under normal conditions but vulnerable to disruption.

Beyond physical infrastructure, institutional arrangements shape resilience outcomes. Governance structures determine how decisions are made, who participates in those decisions, and how conflicts between competing water uses are resolved. Regulatory frameworks establish standards for water quality, environmental protection, and service delivery. Financial mechanisms influence what investments are made and how costs are recovered. Organizations with clear mandates, adequate resources, skilled personnel, and adaptive management cultures are better positioned to respond to emerging challenges than those lacking these attributes.

Social dimensions of resilience deserve equal attention. Communities with strong social networks, local knowledge of water resources, and traditions of collective action often demonstrate remarkable capacity to cope with water-related stresses (Samantara 2023; Chin et al. 2025). Conversely, marginalized populations frequently bear disproportionate burdens when water systems fail, lacking the resources to purchase alternative supplies or the political voice to demand improved services. Equity considerations are thus integral to resilience; a system that serves some populations well while leaving others vulnerable cannot be considered truly resilient.

Ecological health underpins all other dimensions of water resilience. Healthy watersheds regulate water flows, filter pollutants, recharge groundwater, and moderate the extremes of floods and droughts (Tripathi et al. 2025; Kumar et al. 2025b). Degraded ecosystems lose these capacities, transferring the burden of water management entirely to engineered systems that may prove inadequate to the task. Recognizing this, contemporary approaches to water resilience increasingly emphasize the protection and restoration of natural systems as complements to built infrastructure.

The integration of these components distinguishes resilient systems from those that merely appear robust. A city may possess state-of-the-art treatment facilities yet remain vulnerable if the watershed supplying its raw water is degraded, if institutional fragmentation prevents coordinated response to emergencies, or if portions of its population lack access to services. True resilience emerges from the alignment of physical, institutional, social, and ecological elements in mutually reinforcing configurations.

4.2.2 *Threats to Water Resilience*

Contemporary water systems face an array of threats that test their resilience across multiple dimensions. Climate change is perhaps the most pervasive challenge, altering the fundamental hydrological conditions on which water management has been premised. Shifting precipitation patterns redistribute water in space and time, with some regions experiencing increased aridity while others face more intense rainfall events. Temperature increases accelerate evaporation, reduce snowpack, and increase water demand for irrigation and cooling. Sea level rise threatens coastal aquifers and infrastructure with saltwater intrusion and inundation. The extremes of both floods and droughts intensify, straining systems designed for historical variability.

Population growth and urbanization concentrate water demand, stressing available resources. Cities in developing regions are expanding at unprecedented rates, often outpacing the capacity of water infrastructure to keep pace. Informal settlements often lack access to piped water and sanitation, with residents relying on expensive, often unsafe alternatives. Even in developed countries, aging infrastructure requires massive investment for rehabilitation and replacement, competing for limited public resources against other pressing needs. Agricultural water use, which accounts for approximately 70% of global freshwater withdrawals, faces mounting pressures. Irrigated agriculture has expanded dramatically over the past century, often drawing on groundwater reserves that are being depleted far faster than natural recharge can replenish them (Wang et al. 2025; Agarwal et al. 2025). Competition between agricultural and urban water uses intensifies as cities grow, raising difficult questions about allocation priorities and the economic valuation of water.

Pollution degrades water quality, effectively reducing available supplies even where volumetric quantities remain adequate. Industrial discharges, agricultural runoff carrying fertilizers and pesticides, inadequately treated sewage, and emerging contaminants, including pharmaceuticals and microplastics, all compromise water resources. Climate change exacerbates water quality challenges by increasing temperatures, reducing dissolved oxygen levels, concentrating pollutants during low-flow periods, and promoting harmful algal blooms. Infrastructure deterioration poses risks that accumulate gradually but can manifest suddenly in catastrophic failures. Many water systems in developed countries were constructed in the mid-twentieth century and are approaching or exceeding their design lifespans. Deferred maintenance creates backlogs of needed repairs, while the full extent of deterioration often remains hidden until failures occur. Climate change accelerates infrastructure degradation through mechanisms including freeze–thaw cycles, soil movement during droughts, and corrosion from changing water chemistry.

4.2.3 *Measuring and Benchmarking Water Resilience*

Efforts to enhance water resilience require methods to assess current conditions and track progress over time. Various frameworks have emerged to measure water resilience at scales ranging from individual utilities to nations and river basins. These frameworks typically employ indicators spanning the multiple dimensions of resilience, including infrastructure condition, institutional capacity, financial sustainability, environmental health, and social equity. Quantitative metrics provide one avenue for assessment. These may include measures such as the percentage of population with access to safely managed water services, the frequency and duration of service interruptions, water loss rates in distribution systems, the diversity of water sources, the ratio of water withdrawals to available renewable resources, and financial indicators such as cost recovery ratios and investment levels. Such metrics enable comparison across systems and tracking of trends over time, supporting evidence-based management and accountability.

However, quantitative indicators alone cannot capture the full complexity of resilience. Qualitative assessments complement numerical measures by examining factors such as governance effectiveness, stakeholder engagement, adaptive management practices, and organizational culture. Scenario-based assessments test system performance under a range of plausible future conditions, revealing vulnerabilities that may not be apparent under current circumstances. Participatory approaches engage diverse stakeholders in identifying priorities and evaluating trade-offs, ensuring that assessments reflect the values and concerns of affected communities. Benchmarking against peer systems and established standards supports continuous improvement. International initiatives such as the Alliance for Water Stewardship and the International Water Association's water safety planning framework provide reference points against which systems can evaluate their practices. National regulatory frameworks establish minimum standards for service quality and environmental protection. Voluntary certification schemes recognize exemplary performance and create incentives for advancement beyond minimum requirements.

The connection between water resilience assessment and carbon-neutrality goals warrants particular attention. Traditional resilience metrics have not typically incorporated carbon considerations, potentially favouring high-emission solutions that enhance water security in the near term while contributing to climate change that undermines it over longer horizons. Integrating carbon accounting into resilience assessment ensures that the full implications of management choices are visible to decision-makers. This integration supports the identification of solutions that advance both objectives simultaneously, avoiding trade-offs that sacrifice one for the other. Emerging approaches to resilience assessment increasingly embrace systems thinking, recognizing that water systems are embedded within broader social, economic, and ecological contexts. Urban water resilience, for instance, cannot be understood in isolation from land use planning, transportation systems, energy networks, and food supply chains. River basin resilience depends on coordination among multiple jurisdictions and sectors whose decisions collectively determine outcomes. This systems perspective reveals leverage points where interventions can yield benefits across multiple domains, including the water-carbon nexus that this chapter focuses on.

4.3 The Carbon Footprint of Water Systems

Understanding the carbon footprint of water systems requires a comprehensive examination of emissions across the entire water cycle, from source to tap and back again. While water itself is carbon-free, the infrastructure and processes required to deliver safe and reliable water services consume substantial energy and generate significant greenhouse gas emissions. Quantifying these emissions reveals opportunities for reduction and establishes baselines against which progress toward carbon neutrality can be measured.

4.3.1 Energy Consumption in Water Extraction, Treatment, and Distribution

The energy intensity of water supply varies enormously depending on the source, the distance and elevation over which water must be transported, and the treatment required to meet quality standards. Surface water drawn from nearby rivers or lakes that requires only conventional treatment typically represents the least energy-intensive option, with energy requirements often below 0.5 kilowatt-hours per cubic meter (Rashedin 2023). Groundwater extraction requires more energy for pumping, with requirements increasing as water tables decline and wells must reach greater depths. In regions where aquifers have been severely depleted, energy for groundwater pumping can exceed 1.5 kilowatt-hours per cubic meter (Silori et al. 2024; Scanlon et al. 2023; Rizvi et al. 2025).

Figure 4.2 highlights the wide variation in energy requirements (kWh/m^3) for different water sources, treatment processes, distribution methods, and wastewater treatment options. Long-distance water transfers substantially increase energy requirements. Massive conveyance projects that move water across hundreds of kilometres and over mountain ranges rank among the most energy-intensive water-supply options. The California State Water Project, which pumps water from northern California over the Tehachapi Mountains to southern California, consumes approximately 2.5 kilowatt-hours per cubic meter, making it one of the largest single consumers of electricity in the state. Similar projects exist worldwide, representing enormous, embedded energy demands that persist for the operational lifetimes of the infrastructure.

Desalination occupies the high end of the energy intensity spectrum. Seawater reverse osmosis, the dominant desalination technology, typically requires 3–4.5 kilowatt-hours per cubic meter, though technological improvements continue to reduce this figure. Thermal desalination processes consume even more energy. As freshwater sources become increasingly stressed, many water-scarce regions are turning to desalination as a last-resort supply option, with significant implications for energy consumption and carbon emissions. The Middle East, North Africa, and parts of Australia, the United States, and Spain have invested heavily in desalination capacity, creating new demands on energy systems that must be factored into decarbonization planning.

Water treatment for potable use increases energy requirements, with the extent depending on source water quality and the treatment technologies employed (Agarwal et al. 2024). Conventional treatment involving coagulation, flocculation, sedimentation, filtration, and disinfection typically requires between 0.1 and 0.3 kilowatt-hours per cubic meter. Advanced treatment processes, including membrane filtration, ultraviolet disinfection, and advanced oxidation, consume additional energy. Water reuse applications that treat wastewater to potable standards may require energy inputs comparable to those of desalination, though typically lower.

Distribution systems consume energy primarily for pumping to maintain pressure throughout pipe networks and to lift water to elevated storage tanks. The energy

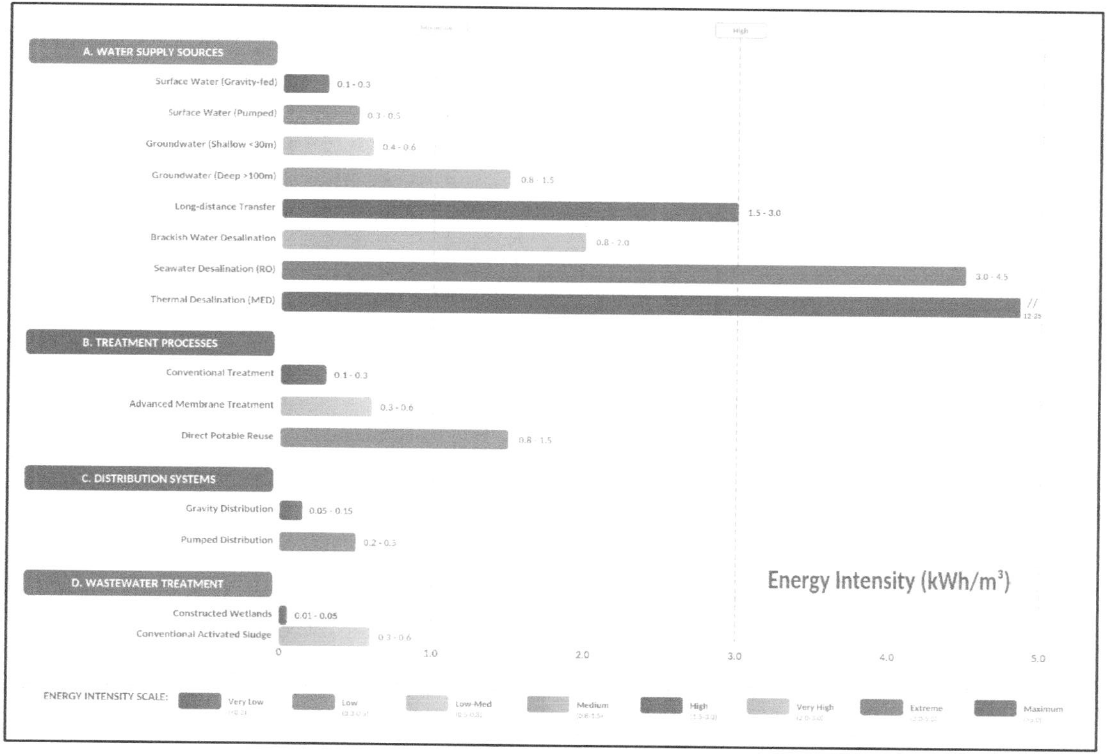

Fig. 4.2 Comparative energy intensity across the urban water cycle. This figure illustrates the wide variation in energy requirements (kWh/m^3) for different water sources, treatment processes, distribution methods, and wastewater treatment options. *Sources* Compiled from IEA World Energy Outlook (2016), Water Research Foundation (2013), EPRI (2013), International Desalination Association (2021), and Plappally and Lienhard (2012)

required depends on system topography, pipe condition, and network configuration. Older systems with deteriorated pipes and suboptimal configurations may consume substantially more energy than well-maintained, efficiently designed networks. Water losses through leakage represent not only wasted water but wasted energy, as the energy invested in extracting, treating, and pumping lost water yields no useful service.

4.3.2 Emissions from Wastewater Management

Wastewater systems generate greenhouse gas emissions through multiple pathways that extend beyond energy consumption (Tian et al. 2022). Collection systems require energy for pumping, particularly in flat terrain where gravity flow is insufficient. Treatment processes consume energy for aeration, mixing, pumping, and solids processing. However, wastewater treatment also releases process emissions of carbon dioxide, methane, and nitrous oxide that are independent of energy use. Methane emissions from wastewater represent a significant and often underappreciated contribution to the sector's carbon footprint (Moore et al. 2023). Anaerobic conditions in sewers, treatment processes, and sludge handling generate methane as microorganisms decompose organic matter in the absence of oxygen. Methane is a potent greenhouse gas with a global warming potential approximately 28–34 times that of carbon dioxide over a hundred-year horizon. Septic systems and pit latrines, which serve billions of people worldwide who lack access to sewered sanitation, generate methane emissions that are difficult to quantify but potentially substantial.

Nitrous oxide emissions from wastewater treatment add another layer of complexity. Biological nitrogen removal processes, which convert ammonia to nitrogen gas through nitrification and denitrification, can generate nitrous oxide as a byproduct. Nitrous oxide has a global warming potential approximately 265–298 times that of carbon dioxide, making even small emissions significant in carbon accounting terms (Song et al. 2024; Yang and Yuan 2023). The magnitude of nitrous oxide emissions varies with treatment process design and operation, creating opportunities for reduction through optimized management.

Sludge management represents both a challenge and an opportunity from a carbon perspective. Landfilling of sludge generates methane emissions as organic matter decomposes anaerobically. Incineration releases carbon dioxide and requires energy input. However, anaerobic digestion of sludge produces biogas, a methane-rich gas that can be used for energy generation, potentially offsetting emissions from other sources. Land application of treated biosolids can sequester carbon in soils while providing agricultural benefits, though careful management is required to avoid other environmental impacts.

4.3.3 Embodied Carbon in Water Infrastructure

Beyond operational emissions, water infrastructure carries a substantial burden of embodied carbon, representing emissions generated during the manufacturing of materials, construction processes, and eventual decommissioning. Concrete and steel, the dominant materials in water infrastructure, are carbon-intensive to produce. Cement manufacturing alone accounts for approximately 80% of global carbon dioxide emissions, while steel production contributes another 7–9%. Large dams, treatment plants, and pipeline networks thus represent significant upfront carbon investments that must be amortized over their operational lifetimes.

Table 4.1, Comparative Energy Intensity and Carbon Emissions Across Water Supply Sources and Processes. The embodied carbon in water infrastructure has received less attention than operational emissions, but its significance is increasingly recognized. Life cycle assessments that account for embodied carbon may yield different conclusions about optimal infrastructure choices than analyses focused solely on operational energy. A nature-based solution with minimal concrete and steel may compare favourably with a conventional engineered alternative, even if operational energy requirements are similar, once embodied carbon is taken into account.

Decisions about infrastructure rehabilitation versus replacement carry embodied carbon implications. Extending the life of existing infrastructure through repair and refurbishment avoids the embodied carbon associated with new construction, though operational efficiency may be compromised. These trade-offs require careful analysis that weighs embodied carbon, operational emissions, service quality, and financial considerations over extended time horizons. The construction phase of infrastructure projects generates additional emissions from the transportation of materials, the operation of construction equipment, and site preparation activities. These emissions vary depending on project scale, location, and construction methods. Innovations in construction practices, including the use of lower-carbon materials, the electrification of construction equipment, and the optimisation of logistics, offer potential to reduce construction-phase emissions.

Circular economy principles applied to water infrastructure can reduce embodied carbon by extending material lifespans and enabling reuse. Modular infrastructure designs facilitate component replacement without wholesale reconstruction. Recycled materials can substitute for virgin inputs in some applications. Design for deconstruction enables recovery of materials at the end of life for use in new projects. These approaches remain relatively novel in the water sector but align with broader sustainability trends and merit expanded application.

Table 4.1 Comparative energy intensity and carbon emissions across water supply sources and processes

Water source/process	Energy intensity (kWh/m^3)	Carbon emissions (kg CO_2e/m^3)	Embodied carbon level	Primary GHG type	Resilience impact	Technical maturity	Key factors affecting emissions	Relative cost implications	Carbon reduction potential
A. Water Supply									
Surface (gravity)	0.1–0.3	0.05–0.15	Low	CO_2	High (reliability)	Established	Proximity to source	Very low	Limited
Surface (pumped)	0.3–0.5	0.15–0.25	Moderate	CO_2	Moderate	Established	Pump efficiency	Low	Moderate
Groundwater (deep)	0.8–1.5	0.40–0.75	Moderate	CO_2	Low (depletion)	Established	Aquifer depth	Moderate-high	High
Long-distance transfer	1.5–3.0	0.75–1.50	Very high	CO_2	Variable	Established	Elevation gains	High	High
B. Desalination									
Seawater RO	3.0–4.5	1.50–2.25	High	CO_2	High (security)	Established	Membrane efficiency	Very high	Very high
Thermal (MED)	12.0–25.0	6.00–12.50	High	CO_2	High (security)	Established	Thermal source	Very high	Very high
C. Treatment									
Conventional	0.1–0.3	0.05–0.15	Moderate	CO_2	Moderate	Established	Chemical dosing	Low	Moderate
Direct potable reuse	0.8–1.5	0.40–0.75	High	CO_2	Very high	Emerging	Barrier requirements	Moderate-high	High

(continued)

Table 4.1 (continued)

Water source/ process	Energy intensity (kWh/m^3)	Carbon emissions (kg CO_2e/m^3)	Embodied carbon level	Primary GHG type	Resilience impact	Technical maturity	Key factors affecting emissions	Relative cost implications	Carbon reduction potential
D. Wastewater									
Activated sludge	0.3–0.6	0.30–0.80	High	N_2O, CO_2	Moderate	Established	Aeration/ process	Moderate	High
Anaerobic	0.05–0.15	−0.20–0.10	Moderate	CH_4	High	Established	Biogas recovery	Low	Very high
Constructed wetlands	0.01–0.05	0.05–0.20	Very low	CH_4, CO_2	High (eco)	Established	Climate/ vegetation	Very low	Moderate

Notes

- Energy intensity values represent typical ranges; actual values vary based on local conditions, technology vintage, and operational practices
- Carbon emissions calculated assuming average grid electricity emission factor of 0.5 kg CO_2e/kWh; actual emissions vary significantly by region
- * Wastewater treatment emissions include direct process emissions (CH_4, N_2O) in addition to energy-related emissions
- ** Negative values indicate net carbon benefit when biogas is recovered and utilized for energy generation
- Carbon Reduction Potential indicates the relative opportunity for emissions reduction through efficiency improvements, renewable energy, or technology substitution

Sources Compiled from International Energy Agency (IEA), Water Research Foundation, and Peer-Reviewed Literature

4.4 Strategies and Innovations for Carbon–Neutral Water Resilience

The transition toward carbon–neutral water systems requires a comprehensive approach that combines proven strategies with emerging innovations. This section examines the pathways available to water managers seeking to reduce emissions while enhancing resilience, spanning efficiency improvements, renewable energy adoption, nature-based solutions, and technological advances. Figure 4.3 summarises four Pillars of Carbon–Neutral Water Management.

4.4.1 Energy Efficiency and Demand Management

The least carbon-intensive energy is that which is never consumed. Energy efficiency improvements represent the foundation of any decarbonization strategy, offering immediate emissions reductions while generating cost savings that can finance further investments. Water utilities worldwide have demonstrated that substantial efficiency gains are achievable through systematic attention to energy use across operations. Pumping systems typically account for the largest share of energy consumption in water utilities, often exceeding 80% of total electricity use. Variable frequency drives that match pump speed to actual demand conditions can reduce pumping energy by 20–50% compared to fixed-speed alternatives. Optimization of pump scheduling to take advantage of time-of-use electricity pricing and to minimize energy losses in the distribution system yields additional savings. Advanced control systems using real-time data and predictive algorithms enable increasingly sophisticated optimization that was previously impractical.

Reducing water losses addresses both water scarcity and energy consumption simultaneously. Every cubic meter of water lost through leakage represents wasted energy invested in extraction, treatment, and distribution. Utilities with high leakage rates, which in some systems exceed 40% of water entering distribution, have enormous potential for efficiency gains. Pressure management, pipe rehabilitation, and active leak detection programs can dramatically reduce losses while improving service reliability. Demand-side management reduces the volume of water requiring treatment and distribution. Water conservation programs, efficiency standards for fixtures and appliances, tiered pricing structures, and public education campaigns all contribute to reducing per capita consumption. Lower demand translates directly to lower energy consumption and emissions, while also reducing stress on water resources and deferring the need for new supply infrastructure.

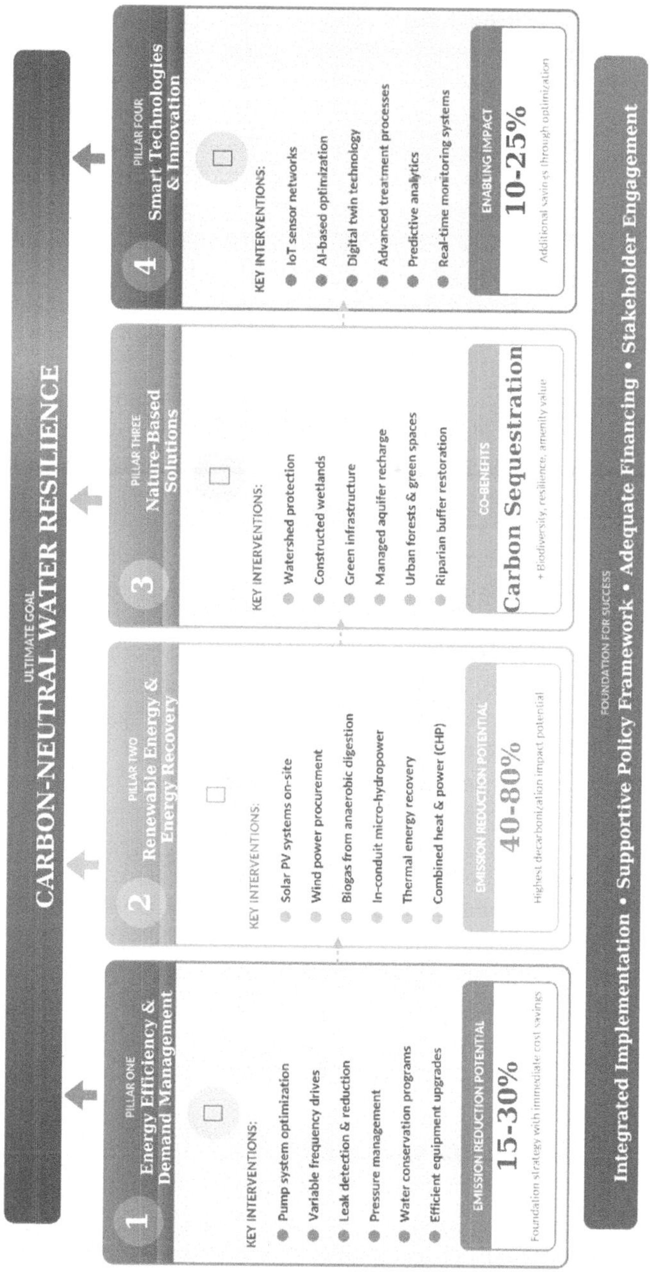

Fig. 4.3 Four pillars of carbon–neutral water management

4.4.2 Renewable Energy Integration and Energy Recovery

Transitioning from fossil fuel-derived electricity to renewable sources offers the most direct pathway to decarbonizing water operations. Many utilities have begun installing solar photovoltaic systems on treatment plant sites, reservoir covers, and other available land. The predictable energy demands of water operations align well with solar generation patterns in many regions, and on-site generation can provide resilience against grid disruptions while hedging against electricity price volatility. Wind power, whether procured through power purchase agreements or direct ownership, provides another avenue for renewable electricity supply. Some utilities have achieved 100% renewable electricity through combinations of on-site generation and off-site procurement. The declining costs of renewable energy have made these transitions increasingly economically attractive, with payback periods shortening as technology costs fall and fossil fuel prices fluctuate.

Hydropower recovery from water supply and distribution systems captures energy that would otherwise be dissipated. Pressure-reducing valves in distribution systems can be replaced with micro-turbines that generate electricity while performing the same pressure regulation function. In-conduit hydropower installations in water transmission pipelines harvest energy from flowing water without the environmental impacts associated with conventional hydroelectric dams. Wastewater treatment offers particularly rich opportunities for energy recovery. Anaerobic digestion of sewage sludge produces biogas containing approximately 60% methane, which can be combusted to generate electricity and heat or upgraded to biomethane for injection into natural gas networks or use as vehicle fuel. Combined heat and power systems using biogas can meet a substantial portion of treatment plant energy needs while reducing methane emissions that would otherwise escape to the atmosphere. Some advanced facilities have achieved energy neutrality or even become net energy producers through aggressive implementation of efficiency measures and energy recovery.

Thermal energy recovery from wastewater represents an emerging opportunity. Sewage maintains relatively constant temperatures year-round, making it an effective source for heat pump systems that can provide heating and cooling for treatment plant buildings or adjacent developments. District heating systems in some European cities utilize wastewater heat on a larger scale, demonstrating the potential for integration between water and energy systems.

4.4.3 Nature-Based Solutions and Green Infrastructure

Nature-based solutions harness ecological processes to deliver water management functions while providing carbon benefits that engineered alternatives cannot match. Watershed protection preserves forests and wetlands that regulate water flows, filter pollutants, and sequester carbon in biomass and soils. The carbon stored in intact ecosystems would be released if those systems were degraded, making protection a form of avoided emissions in addition to an ongoing sequestration benefit. Constructed wetlands for wastewater treatment combine effective pollutant removal with substantially lower energy requirements than conventional mechanical treatment. These systems rely on natural processes including microbial activity, plant uptake, and sedimentation to achieve treatment objectives. While requiring more land area than conventional facilities, constructed wetlands offer advantages including lower operating costs, reduced chemical inputs, habitat creation, and aesthetic values that mechanical systems cannot provide.

Urban green infrastructure manages stormwater through distributed interventions that mimic natural hydrology. Rain gardens, bioswales, permeable pavements, and green roofs capture and infiltrate rainfall, reducing volumes requiring conveyance and treatment in centralized systems. Urban trees and vegetation provide evaporative cooling that reduces both stormwater runoff and urban heat island effects, with associated carbon sequestration in woody biomass. These distributed approaches enhance resilience by reducing dependence on centralized infrastructure that may be vulnerable to extreme events. Managed aquifer recharge stores surplus water underground during wet periods for recovery during dry periods, providing a low-energy alternative to surface reservoirs. The energy required for injection and recovery is typically modest compared to long-distance surface water transfers or desalination. Aquifer storage also avoids evaporative losses that can claim significant fractions of surface reservoir contents in arid climates.

4.4.4 Smart Systems and Advanced Technologies

Digital technologies are transforming water management in ways that support both efficiency and resilience. Smart water networks incorporate sensors throughout infrastructure to monitor flows, pressures, and water quality in real time. Advanced analytics applied to sensor data enable rapid leak detection, predictive maintenance, and optimized operations that reduce energy consumption while improving service quality. Artificial intelligence and machine learning applications are moving from research into operational deployment. These technologies can optimize complex systems in ways that exceed human analytical capacity, identifying patterns and opportunities that would otherwise remain hidden. Energy optimization, demand forecasting, and treatment process control all benefit from intelligent automation that continuously learns and adapts.

Advanced treatment technologies offer potential for reducing energy intensity while meeting increasingly stringent quality requirements. Membrane bioreactors combine biological treatment with membrane filtration in compact configurations that can reduce footprint and energy consumption compared to conventional alternatives. Anaerobic treatment processes generate rather than consume energy, though their applicability depends on wastewater characteristics and effluent requirements. Decentralized and hybrid systems distribute treatment and reuse functions closer to points of generation and use, reducing energy for conveyance while enhancing resilience through redundancy. Building-scale and neighborhood-scale water recycling systems can provide fit-for-purpose water quality for non-potable applications including irrigation, toilet flushing, and cooling. These systems reduce demands on centralized infrastructure while enabling water reuse that might not be feasible at larger scales.

Carbon capture and utilization technologies, while still emerging, offer potential for transforming wastewater treatment from a carbon source to a carbon sink. Algae cultivation systems can capture carbon dioxide while producing biomass for energy or other uses. Mineralization processes can sequester carbon in stable forms while producing materials with commercial value. These approaches remain largely at pilot scale but merit continued development as components of long-term decarbonization strategies.

4.5 Policy Frameworks, Governance, and Overcoming Barriers

The technical and financial feasibility of carbon–neutral water resilience has been established through numerous demonstrations worldwide. However, translating these possibilities into widespread practice requires supportive policy frameworks, effective governance arrangements, and deliberate strategies to overcome persistent barriers. This section examines the enabling conditions necessary for accelerating the transition and the obstacles that must be navigated along the way.

4.5.1 Integrating Water and Climate Policies

Historically, water and climate policies have developed along separate tracks, with different institutions, legal frameworks, and professional communities addressing each domain. Water management has been governed primarily by regulations focused on public health protection, environmental quality, and resource allocation, with limited attention to greenhouse gas emissions. Climate policies, conversely, have concentrated on energy and transportation sectors while largely overlooking the water sector's contribution to emissions and its vulnerability to climate impacts. This fragmentation impedes integrated solutions that could advance both agendas simultaneously. Progressive jurisdictions are beginning to bridge this divide through policies that explicitly connect water and climate objectives. Mandates requiring water utilities to measure and report greenhouse gas emissions establish accountability and enable tracking of progress. Emissions reduction targets specific to the water sector create clear expectations and drive innovation. Integration of water considerations into national climate strategies and inclusion of climate considerations in water planning documents ensure that decision-makers in each domain account for implications in the other.

Carbon pricing mechanisms, whether through taxes or cap-and-trade systems, create economic incentives for emissions reduction across all sectors including water. When utilities face costs for their emissions, investments in efficiency, renewable energy, and low-carbon technologies become more financially attractive. However, the effectiveness of carbon pricing depends on price levels sufficient to influence behavior and on complementary policies that address barriers beyond simple economics.

4.5.2 Financing Carbon–Neutral Water Infrastructure

The capital requirements for transitioning to carbon–neutral water systems are substantial, demanding innovative financing approaches that can mobilize resources at the necessary scale. Traditional utility financing through ratepayer revenues and municipal bonds remains essential but may prove insufficient for the accelerated investment required. Blended finance approaches that combine public funding, private investment, and concessional capital from development institutions can bridge financing gaps while managing risks appropriately.

Green bonds have emerged as significant instruments for financing climate-related water infrastructure. These bonds, whose proceeds are dedicated to environmentally beneficial projects, have attracted growing investor interest as financial markets increasingly incorporate sustainability considerations. Water utilities and municipalities have successfully issued green bonds for projects including renewable energy installations, efficiency improvements, and nature-based solutions. The green bond market provides access to capital while signaling organizational commitment to sustainability.

Climate finance mechanisms established under international agreements offer additional resources for developing countries. The Green Climate Fund and other multilateral facilities provide grants and concessional loans for climate mitigation and adaptation projects, including water sector investments that reduce emissions or enhance resilience. Accessing these resources requires project development capacity and alignment with fund priorities, representing both opportunity and challenge for water institutions in eligible countries.

4.5.3 Technical, Financial, and Institutional Barriers

Despite the availability of technologies and financing mechanisms, numerous barriers impede progress toward carbon–neutral water resilience. Technical barriers include limitations in data availability for emissions accounting, uncertainty about performance of emerging technologies at scale, and workforce capacity gaps in areas including data analytics, renewable energy, and nature-based solutions. Many utilities lack the monitoring systems necessary to establish emissions baselines and track reduction progress, hindering both management and accountability. Financial barriers extend beyond capital availability to encompass misaligned incentives and cost recovery challenges. Utilities operating under rate structures that reward increased water sales face disincentives for conservation programs that would reduce both demand and emissions. Upfront costs of efficient equipment and renewable energy systems may deter investment even when lifecycle economics are favorable. Benefits of emissions reduction accrue broadly to society while costs are borne by utilities and their ratepayers, creating a mismatch that policy interventions must address.

Institutional barriers often prove most intractable. Fragmented responsibilities across multiple agencies complicate coordinated action on water and climate challenges. Short political cycles and leadership turnover can disrupt long-term planning and implementation. Risk aversion within utilities may favor familiar approaches over innovative alternatives with uncertain outcomes. Regulatory frameworks designed for conventional infrastructure may inadequately accommodate nature-based solutions or decentralized systems. Procurement processes may prioritize lowest upfront cost over lifecycle performance and emissions implications.

Capacity constraints affect utilities of all sizes but pose particular challenges for smaller systems lacking specialized staff and resources. These utilities, which collectively serve substantial populations, may struggle to access technical assistance, navigate complex financing mechanisms, or implement sophisticated management approaches. Addressing their needs requires targeted support including technical assistance programs, collaborative arrangements that enable resource sharing, and simplified pathways to adopt proven solutions.

4.5.4 Strategies for Accelerating Progress

Overcoming these barriers requires deliberate strategies operating at multiple levels. Knowledge sharing and peer learning networks enable utilities to benefit from others' experiences, reducing the costs and risks of innovation. Professional associations, research institutions, and international organizations facilitate such exchange, though more systematic efforts could accelerate diffusion of best practices. Capacity building investments strengthen the human and institutional foundations necessary for sustained progress. Training programs for utility staff, technical assistance for planning and implementation, and support for organizational development all contribute to enhanced capacity. These investments yield returns over extended periods as strengthened institutions continue delivering benefits long after initial support concludes.

Regulatory reform can remove barriers while creating positive incentives for carbon–neutral approaches. Performance-based regulations that focus on outcomes rather than prescribing specific technologies provide flexibility for innovation. Streamlined permitting for beneficial projects reduces transaction costs and delays. Requirements for lifecycle cost analysis in infrastructure decisions ensure that carbon implications receive appropriate consideration. Public engagement builds support for investments and behavioral changes necessary for carbon–neutral water resilience. Transparent communication about the connections between water services, energy consumption, and climate change helps ratepayers and citizens understand the rationale for utility actions. Participatory processes that involve stakeholders in planning and decision-making generate ownership and legitimacy for resulting choices.

4.6 Conclusion

The journey toward carbon–neutral water resilience represents one of the defining challenges of our era, demanding a fundamental reimagining of how societies manage their most essential resource. Throughout this chapter, we have examined the intricate connections between water systems and carbon emissions, revealing a relationship that operates through multiple pathways and across multiple scales. The water-carbon nexus is not merely an academic construct but a practical reality that shapes the sustainability of communities worldwide.

The evidence presented demonstrates that water systems contribute meaningfully to global greenhouse gas emissions through energy-intensive extraction, treatment, and distribution processes; through direct process emissions from wastewater management; and through embodied carbon in infrastructure. These emissions, while modest compared to sectors such as energy and transportation, nonetheless represent a significant and often overlooked component of humanity's carbon footprint. More critically, the carbon emissions from water infrastructure contribute to climate change that in turn threatens the water resources upon which human societies depend, creating a feedback loop that demands urgent attention.

Yet the picture that emerges from this analysis is far from hopeless. The pathways to carbon–neutral water resilience are numerous and increasingly well-understood. Energy efficiency improvements offer immediate opportunities for emissions reduction while generating cost savings. Renewable energy integration enables decarbonization of electricity-dependent operations. Nature-based solutions provide multifunctional benefits that engineered alternatives cannot match, simultaneously enhancing resilience, sequestering carbon, and delivering co-benefits for biodiversity and human wellbeing. Emerging technologies continue expanding the toolkit available to water managers, from smart systems that optimize operations to advanced treatment processes that recover resources and energy from wastewater.

The barriers to progress, while real, are not insurmountable. Technical challenges yield to sustained research and development. Financial constraints can be addressed through innovative mechanisms that mobilize capital at necessary scales. Institutional obstacles, though often most persistent, respond to deliberate reform efforts and capacity building investments. The examples of utilities and communities that have achieved or approached carbon neutrality demonstrate that ambitious goals are attainable with committed leadership and supportive policy environments.

Looking forward, the imperative for action will only intensify. Climate change impacts on water resources are accelerating, making resilience investments increasingly urgent. Global commitments to emissions reduction demand contributions from all sectors, including water. Growing public awareness of environmental challenges creates both expectations and opportunities for water institutions to demonstrate leadership. The choices made in the coming years about water infrastructure and management practices will shape outcomes for decades, making it essential that those choices reflect full consideration of carbon implications alongside traditional concerns of reliability, quality, and cost.

The integration of water resilience and carbon neutrality objectives represents not a burden but an opportunity. Solutions that advance both goals simultaneously offer superior value compared to approaches that address each in isolation. The transition to carbon–neutral water systems can enhance rather than compromise service quality, reduce rather than increase long-term costs, and strengthen rather than weaken community resilience. Realizing this potential requires vision, commitment, and collaboration across the many institutions and stakeholders whose decisions collectively determine outcomes in the water sector.

As we conclude this chapter, it is worth reflecting on the broader significance of the water-carbon nexus for sustainable development. Water and climate are not merely environmental concerns but foundations of human welfare, economic prosperity, and social stability. Communities that successfully navigate the transition to carbon–neutral water resilience will be better positioned to thrive in an uncertain future, while those that fail to act will face compounding challenges as climate impacts intensify and infrastructure ages. The stakes could hardly be higher, but neither could the opportunities for those willing to embrace the integrated approaches this moment demands.

References

Agarwal V, Kumar M, Panday DP, Zang J, Munoz-Arriola F (2024) Unlocking the potential of remote sensing for arsenic contamination detection and management: challenges and perspectives. Curr Opin Environ Sci Health 42:100578

Agarwal V, Kumar M, Dogra K, Mejia-Avendaño S (2025) Isomers of perfluorooctanesulfonate exhibit preferential infiltration and contrasting ionic associations between surface water and groundwater. J Hazard Mater 494:138445

Cardoso BJ, Gomes Á, Gaspar AR (2023) Barriers and drivers to energy efficiency in the Portuguese water sector: survey analysis. Appl Energy 333:120630

Chen L, Msigwa G, Yang M, Osman AI, Fawzy S, Rooney DW, Yap PS (2022) Strategies to achieve a carbon neutral society: a review. Environ Chem Lett 20(4):2277–2310

Chin SR, Ramnath SO, Khan K, Elibox W (2025) Learning resilience: gendered adaptation and community-led water management in a resource-stressed village in a small island developing state. J Work-Appl Manag 1–14

Chowdhury R, Agarwal V (2024) Mathematical models and dynamic global warming potential calculation for estimating the role of organic amendment in net-zero goal achievement. Energies 17(19):4819

Huang Y, Zhang J, Ren Z, Xiang W, Sifat I, Zhang W et al (2023) Next generation decentralized water systems: a water-energy-infrastructure-human nexus (WEIHN) approach. Environ Sci Water Res Technol 9(10):2446–2471

Jiang Q, Bhattarai N, Pahlow M, Xu Z (2022) Environmental sustainability and footprints of global aquaculture. Resour Conserv Recycl 180:106183

Kumar M, Panday DP, Bhagat C, Herbha N, Agarwal V (2023) Demystifying the decadal shift in the extent of groundwater in the coastal aquifers of Gujarat, India: a case of reduced extent but increased magnitude of seawater intrusion. Sci Total Environ 898:165451

Kumar M, Dash S, Mahlknecht J, Kolok A, Dogra S, Kuroda K et al (2025a) Understanding the pathways, pollution and potential solutions pertaining to pesticides: circular engineering for persistent chemicals. Curr Opin Environ Sci Health 100638

Kumar M, Gupta P, Dogra S, Sarkar D, Mora A, Ornelas-Soto N, Mahlknecht J (2025b) Propensity and repercussion of microplastics in the soil-water-urban continuum. J Contamin Hydrol 104663

Kyriakopoulos GL, Sebos I (2023) Enhancing climate neutrality and resilience through coordinated climate action: review of the synergies between mitigation and adaptation actions. Climate 11(5):105

Lehtoranta S, Laukka V, Silvennoinen K (2025) Climate change impacts of municipal water sector and mitigation pathways: a national scale analysis and perspectives to carbon neutrality. J Environ Manage 373:123732

Li L, Xu H, Zhang Q, Zhan Z, Liang X, Xing J (2024) Estimation methods of wetland carbon sink and factors influencing wetland carbon cycle: a review. Carbon Res 3(1):50

Meinshausen M, Lewis J, McGlade C, Gütschow J, Nicholls Z, Burdon R et al (2022) Realization of Paris agreement pledges may limit warming just below 2 C. Nature 604(7905):304–309

Moore DP, Li NP, Wendt LP, Castañeda SR, Falinski MM, Zhu JJ et al (2023) Underestimation of sector-wide methane emissions from United States wastewater treatment. Environ Sci Technol 57(10):4082–4090

Panday DP, Kumari A, Kumar M (2025) Alkalinity-salinity-sustainability: decadal groundwater trends and its impact on agricultural water quality in the Indian Peninsula. Sci Total Environ 978:179459. https://doi.org/10.1016/j.scitotenv.2025.179459

Panday DP, Agarwal V, Silori R (2026) From regional disputes to global solutions: a fairness-based game-theoretic model for sustainable river water allocation. Water Resour Manage 40(1):11

Qin J, Duan W, Zou S, Chen Y, Huang W, Rosa L (2024) Global energy use and carbon emissions from irrigated agriculture. Nat Commun 15(1):3084

Rashedin M (2023) Investigating factors affecting energy consumption in rural alaskan water treatment and distribution systems, and exploring energy-saving strategies for wastewater treatment in cold climates. University of Alaska Fairbanks

Rizvi SF, Izhar SK, Shahab A, Kuddus M, Roohi R (2025) Comprehensive review on groundwater exploitation, depletion, and its management: sustainable approach toward agricultural productivity in India. Environ Qual Manage 34(4):e70092

Rothausen SG, Conway D (2011) Greenhouse-gas emissions from energy use in the water sector. Nat Clim Chang 1(4):210–219

Samantara A (2023) Community-based water management in urban informal spaces: capabilities and collective action. Doctoral dissertation

Scanlon BR, Fakhreddine S, Rateb A, de Graaf I, Famiglietti J, Gleeson T et al (2023) Global water resources and the role of groundwater in a resilient water future. Nat Rev Earth Environ 4(2):87–101

Silori R, Agarwal V, Panday DP, Giri BS, Khandelwal M, Garg M, Shrivastava V (2024) Assessing groundwater quality in the rapidly industrialising Selaqui Area of Dehradun, India: implications for sustainable water management. Water Conserv Sci Eng 9(2):89

Singh BVR, Batar AK, Agarwal V, Sen A, Kulhari K (2025) Forest fragmentation and human-wildlife conflict: assessing the impact of land use land cover change in Ranthambhore Tiger Reserve, India. Environ Res Commun 7(6):061006

Sinha SK, Davis C, Gardoni P, Babbar-Sebens M, Stuhr M, Huston D et al (2023) Water sector infrastructure systems resilience: a social–ecological–technical system-of-systems and whole-life approach. Cambridge Prisms Water 1:e4

Song C, Zhu JJ, Willis JL, Moore DP, Zondlo MA, Ren ZJ (2024) Oversimplification and misestimation of nitrous oxide emissions from wastewater treatment plants. Nat Sustain 7(10):1348–1358

Sushma N, Suresh HN, Lakshmi JM, Srinivasu PN, Bhoi AK, Barsocchi P (2023) A unified metering system deployed for water and energy monitoring in smart city. IEEE Access 11:80429–80447

Tian Y, Liu S, Guo Z, Wu N, Liang J, Zhao R et al (2022) Insight into greenhouse gases emissions and energy consumption of different full-scale wastewater treatment plants via ECAM tool. Int J Environ Res Public Health 19(20):13387

Tripathi PM, Agarwal V, Mishra V, Singh BVR (2025) Options and strategies for managing droughts: case studies from agro-ecological regions. In: Water sustainability and hydrological extremes. Elsevier, pp 315–331

Wang W, Yang K, Niu Q, Han Z, Zhang J, Agarwal V (2025) Development of in-situ permeability testing system for low-permeability sandstone-type uranium deposits. Adv Geo-Energy Res 18(1):69–83

Yang R, Yuan L (2023) Generation, emission reduction/utilization, and challenges of greenhouse gas nitrous oxide in wastewater treatment plants—a review. J Water Process Eng 53:103871

Ye S, Pei L, He L, Xie L, Zhao G, Yuan H et al (2022) Wetlands in China: evolution, carbon sequestrations and services, threats, and preservation/restoration. Water 14(7):1152

Yue W, Wang S, Xu M, Rong Q, Xu C, Su M (2023) The potential of mitigating greenhouse gas emissions from urban domestic water systems in highly urbanized areas. J Clean Prod 382:135206

Chapter 5
Summary and Future Directions

Abstract This study synthesizes the central thesis that sustainable water management requires an integrated treatment of climate forcing, anthropogenic pressures, water quantity variability, and water quality constraints. It highlights how hydroclimatic extremes and non-stationary human impacts jointly reshape hydrological regimes, creating coupled risks to water security, ecosystems, and public health. The study consolidates key insights on the quality–quantity nexus, emphasizing that degraded quality functions as effective scarcity by reducing usable supply and increasing treatment and governance burdens. It then outlines a forward roadmap built on four pillars: strengthened monitoring through remote sensing and storage diagnostics; climate-aware and interpretable predictive modelling for contaminant forecasting; uncertainty quantification as a core design objective for decision support; and resilience strategies aligned with carbon-neutral pathways. Finally, it underscores adaptive governance as the enabling condition for translating integrated evidence into robust policies and investments under deep uncertainty and accelerating change.

Keywords Climate–human pressures · Water quality–quantity nexus · Predictive modelling · Uncertainty-aware decision support · Carbon-neutral water resilience

5.1 Integrated Synthesis of Climate, Anthropogenic Pressures, and Water System Sustainability

Water sustainability is shaped by the combined effects of climate variability/change and anthropogenic pressures, which jointly alter water availability, reliability, and safety. This book frames water systems as coupled socio-hydrological entities in which climate-driven shifts in precipitation, temperature, and extremes interact with human-driven changes such as land-use transformation, urban growth, groundwater abstraction, and pollutant loading. A central synthesis is that sustainability cannot be evaluated through water quantity or water quality alone; it must be assessed through their co-evolution under combined natural and human forcing. Hydrometeorological extremes play a disproportionate role in sustainability outcomes. Droughts

M. Kumar et al., *Bridging Water Quality and Quantity in the Anthropocene*, SpringerBriefs in Water Science and Technology,
https://doi.org/10.1007/978-3-032-25655-3_5

reduce recharge and flows while increasing dependence on groundwater and intensifying concentration effects; floods can increase availability but often degrade quality through mobilization and infrastructure overload. These risks are increasingly compounded when hazards co-occur or cascade, challenging stationary assumptions and single-hazard design approaches (Sillmann et al. 2017; Zscheischler et al. 2020).

Standardized climate diagnostics translate complex variability into decision-relevant signals that support early warning. SPEI captures warming-driven evaporative demand and therefore remains informative even where precipitation trends are uncertain (Vicente-Serrano et al. 2010), while the Standardized Precipitation Temperature Index further reflects joint temperature–precipitation controls on drought stress (Ali et al. 2017). In addition, teleconnections and large-scale climate modes can extend lead time for drought and flood preparedness when applied in a regionally appropriate way (Nalley et al. 2019; Panday and Kumar 2024). Figure 5.1 summarizes the conceptual framework developed across the entire work. Climate drivers, extremes, and anthropogenic pressures jointly influence water quantity (variability and storage) and water quality (contamination and safety). Their interaction determines hydrological and ecosystem risks, which are addressed through five integrated response pillars: remote sensing and diagnostics, predictive modelling, uncertainty assessment, carbon-neutral resilience, and adaptive governance.

Anthropogenic pressures introduce non-stationarity by modifying runoff generation, infiltration, storage, and connectivity, thereby altering how climate anomalies translate into hydrological and water-quality responses. As a result, sustainability assessments must be dynamic and context-specific rather than assuming historical relationships will persist (Shah et al. 2021). Because uncertainty permeates climate projections, hydrological response, and water quality behaviour, it must be treated as a first-order management variable, with transparent evaluation and interpretation guiding decision support (Beck 1987; Legates and McCabe 1999; Tripathi et al. 2025).

5.2 Interlinkages Amongst Water Quality and Quantity—Humanity and Sustainability

Water security for humanity is increasingly constrained not only by the volume of water available but by the fraction of that water that is safe and usable. The synthesis developed in this book shows that water quantity and water quality are deeply interdependent: changes in hydrological regime affect contaminant concentrations, mobility, and exposure pathways, while quality deterioration reduces effective supply and increases the cost and complexity of meeting human and ecosystem needs. The study further positioned these interlinkages as fundamental to sustainability, arguing that planning based only on allocation or only on contamination

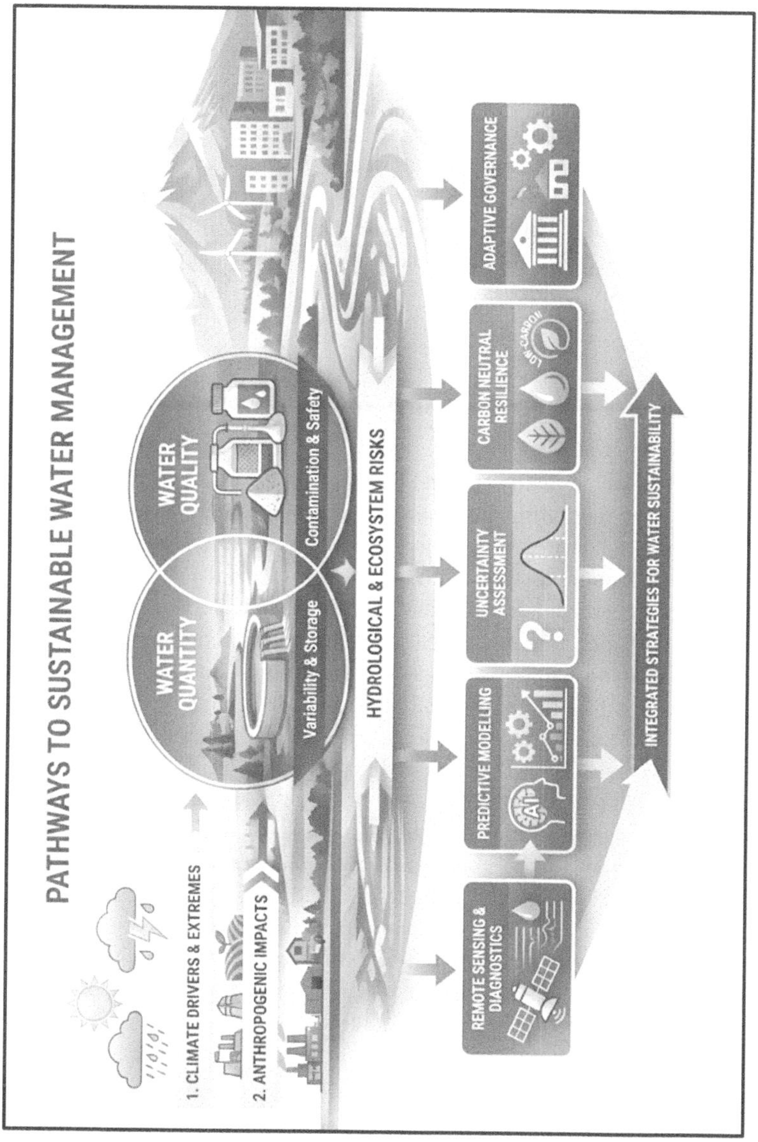

Fig. 5.1 Pathways to sustainable water management. Climate drivers, extremes, and anthropogenic pressures act as primary drivers that jointly shape water quantity (variability and storage) and water quality (contamination and safety). Their coupled behaviour generates hydrological and ecosystem risks

control is structurally incomplete. A core interlinkage arises from the physical principle that many contaminants respond to hydrological variability through dilution, concentration, and residence time effects.

During low-flow or drought periods, reduced dilution and longer water residence times often increase the concentration of dissolved constituents and can intensify exposure risks when communities rely on diminished sources. Conversely, during high-flow events, rapid transport, runoff-driven loading, and sediment mobilization can cause acute, but sometimes short-lived, quality degradation. These patterns become particularly consequential when water systems are pushed toward thresholds: droughts can force reliance on deeper or marginal aquifers, while floods can overwhelm treatment and distribution systems. Such dynamics reinforce that "quantity shocks" frequently become "quality shocks," and vice versa, so integrated assessment is essential for sustainability under climatic and anthropogenic stress (Sillmann et al. 2017; Zscheischler et al. 2020).

5.2.1 Water Quality Constraints as Effective Water Scarcity

In practical terms, poor water quality behaves like scarcity by reducing usable supply, restricting options, and increasing treatment requirements. Public health protection depends on compliance with drinking-water standards and risk-based guidelines, which define acceptable concentrations for chemical and microbial hazards. When water sources exceed guideline thresholds, the system experiences "functional scarcity" even if total water volumes appear adequate, because the available water cannot safely meet demand without additional intervention. This concept is central to sustainability: safeguarding quantity without maintaining quality can still yield insecure outcomes, while protecting quality can preserve effective supply and reduce the need for expensive substitution (World Health Organization 2004).

This interdependence is particularly evident in groundwater-dominated regions, where abstraction increases during drought, and quality challenges can expand concurrently. Even when the book's emphasis is conceptual rather than site-specific, the broader evidence base supports that geogenic contaminants such as arsenic, fluoride, and uranium remain major concerns in many aquifer systems, affecting drinking-water suitability and long-term human health risk. Sustainability planning, therefore, requires managing groundwater as both a quantity reservoir and a quality-sensitive resource, rather than treating groundwater as a fail-safe buffer during surface-water scarcity (CGWB 2018; Vengosh et al. 2022).

5.2.2 Human Health and Exposure Pathways in Coupled Systems

Water quality–quantity linkages manifest most directly through human exposure. Drought-related reductions in supply can shift communities toward alternative sources of uncertain quality, increase reliance on stored water with a higher risk of contamination, or intensify pumping from aquifers with elevated levels of geogenic constituents. Flood-related disruptions can contaminate drinking-water infrastructure and increase pathogen transport, creating acute risks that compound existing chemical exposures. Because these impacts depend on both hydrology and infrastructure, the same climate anomaly can produce very different outcomes depending on governance capacity, monitoring coverage, and treatment resilience.

The sustainability perspective advanced in this book emphasizes that health risk is not solely a function of contaminant concentration; it is also shaped by the availability of safe alternatives, the reliability of distribution systems, and the capacity to manage variability. This framing aligns with drinking-water risk management approaches that prioritize preventative strategies, systematic monitoring, and protective thresholds. Standards and guidelines provide a necessary baseline, but in non-stationary systems, the practical challenge is maintaining compliance under shifting hydroclimatic conditions and anthropogenic pressures (World Health Organization 2004).

5.2.3 Climate Variability, Drought Indices, and Quantity–Quality Coupling

Climate indices provide a structured means of linking climatic variability to coupled water risks. Drought indices such as SPEI incorporate both precipitation and evapotranspiration, making them sensitive to warming-driven increases in atmospheric demand that can intensify drought impacts even without a strong precipitation decline. Because drought affects recharge, surface-water flows, and storage dynamics, it also modifies the context in which contaminants behave—shifting dilution capacity, redox conditions, and residence times in aquifers. The synthesis across the study suggests that drought monitoring should therefore be interpreted as a proxy indicator for periods of enhanced vulnerability to water quality deterioration, especially where water systems are already stressed (Vicente-Serrano et al. 2010; Ali et al. 2017).

Moreover, teleconnection-informed diagnostics can extend lead times for identifying risk windows. When large-scale climate modes signal likely precipitation anomalies, managers can anticipate periods when low flows and reduced recharge may coincide with increased concentrations of dissolved constituents or heightened reliance on groundwater. While teleconnections are not universally strong, the literature synthesized in this book supports their relevance for understanding hydroclimatic variability and extremes at basin scales, which, in turn, indirectly affect water quality through the quantity pathways that govern dilution and transport (Nalley et al. 2019; Panday et al. 2024).

5.2.4 *Governance, Uncertainty, and Integrated Decision-Making*

A key conclusion is that governance structures must be designed to manage **coupled risks** rather than isolated objectives. Water allocation decisions influence water quality by shaping flow regimes, dilution capacity, and residence times. Conversely, contamination constraints influence allocation by limiting safe sources and increasing reliance on alternative supplies. Integrated water resource management, therefore, requires coordinated institutions, shared data infrastructures, and decision-support tools that can evaluate trade-offs across sectors, including domestic use, agriculture, industry, and ecosystem requirements.

Uncertainty compounds these governance challenges. Climate variability introduces uncertainty into future water availability; model uncertainty affects predictions of hydrological response; and water quality uncertainty arises from heterogeneous sources and evolving contaminant profiles. Classic analyses of uncertainty in water quality modelling highlight that prediction should be accompanied by explicit recognition of uncertainty sources and their implications for decision-making (Beck 1987). Similarly, careful model evaluation is required to avoid misleading conclusions based on inappropriate performance metrics or overfitting, especially when models are used to support policy or infrastructure choices (Legates and McCabe 1999). In sustainability contexts, uncertainty strengthens the case for adaptive management approaches that can be revised as new information becomes available.

The integrated framing advanced in this book implies that sustainability pathways must simultaneously protect water quantity and water quality while remaining robust to variability, extremes, and anthropogenic change. This requires shifting from a narrow objective of ‘more water’ toward ‘more safe water’, supported by monitoring systems that capture both quality and quantity signals, and by policies that reduce exposure risks during climate extremes. The interlinkages synthesized in the study also justify the book’s subsequent focus: predictive modelling is needed because coupled systems are nonlinear and increasingly non-stationary, while resilience strategies with carbon-neutral emphasis are needed because traditional supply augmentation can be energy-intensive and may not address quality constraints.

5.3 Predictive Modelling for Contaminant Forecasting

The coupled pressures between water systems make clear that sustainable water management increasingly depends on the ability to anticipate contamination risks rather than only respond to them. The study focused on predictive modelling as a bridge between hydroclimatic variability, human pressures, and actionable water-quality forecasting. The synthesis developed in this book is that predictive models become most useful when they connect quantity-linked drivers—such as precipitation anomalies, drought severity, streamflow variability, and storage dynamics—to water quality outcomes, including contaminant concentrations, risk indices, or water quality indices. In complex real-world systems where contaminants respond nonlinearly to hydrological forcing and anthropogenic loading, predictive modelling offers a practical pathway to early warning, resource prioritization, and risk-informed allocation.

A first principle emphasized in the book is that predictive modelling must respect the coupled structure of water systems. A single variable rarely explains contaminant occurrence; instead, it emerges from interacting controls such as recharge and residence time, redox conditions, source loading, and hydraulic connectivity. Climate indices and teleconnection signals provide additional predictive information by summarizing large-scale drivers of wet and dry regimes that influence dilution capacity and transport pathways. The goal, therefore, is not to treat modelling as a purely statistical exercise, but to integrate climate–hydrology indicators into data-driven frameworks that can generalize beyond narrow calibration windows. Such climate-informed predictive framing is particularly important under non-stationarity, where historical contaminant–flow relationships may drift as climate extremes intensify and human pressures evolve (Sillmann et al. 2017).

5.3.1 *Data-Driven Approaches and the Role of Artificial Intelligence*

Artificial intelligence and machine learning methods have increasingly been applied to water forecasting problems, including rainfall and hydrological prediction, and more recently to water quality forecasting. These approaches are well-suited to environmental systems because they can capture nonlinear relationships and interactions among predictors that are difficult to represent with simple empirical models. Earlier demonstrations of neural network approaches for rainfall prediction illustrate the broader potential of AI to extract predictive structure from hydroclimatic data, particularly when the system exhibits complex dependencies and delayed responses (Abbot and Marohasy 2012). Building on this paradigm, contemporary water quality prediction studies have applied AI methods to forecast quality variables and indices, supporting the claim that data-driven frameworks can contribute meaningfully to anticipatory management if trained, validated, and interpreted responsibly (Aldhyani et al. 2020).

Beyond general AI applications, this study emphasized methods such as Random Forests and hybrid machine learning frameworks because of their flexibility and relatively strong performance across heterogeneous environmental datasets. Hybrid algorithms that combine multiple modelling elements have shown promise for improving the prediction of composite water quality indices, reinforcing the idea that multivariable, integrated targets are often more relevant to management than single-parameter predictions (Bui et al. 2020). The synthesis from this book is that model selection should be guided by the decision context: where explainability and robustness are required for governance uptake, models that provide variable importance and stable generalization may be preferable to highly complex architectures.

5.3.2 *Climate Indices and Teleconnections as Predictors in Contamination Forecasting*

A major contribution of this book is the emphasis on incorporating climate indices into contaminant forecasting workflows. Climate indices can act as compressed, decision-relevant predictors of hydrological conditions that influence contaminant processes. For example, multi-scalar drought indices can indicate sustained deficits that reduce recharge and increase the relative influence of groundwater pumping, while also signaling the likelihood of concentration effects in surface waters due to reduced flows. The standardized precipitation–evapotranspiration index is particularly relevant in warming climates because it explicitly accounts for evaporative demand, thereby capturing drought-intensification pathways not captured by precipitation-only metrics (Vicente-Serrano et al. 2010). Likewise, indices that combine precipitation and temperature provide alternative pathways for representing hydroclimatic stress, potentially strengthening forecasting in regions where temperature-driven demand dominates variability (Ali et al. 2017).

The usefulness of such predictors increases when large-scale climate signals further inform forecasting systems. Reservoir inflow forecasting studies have shown that climate indices can improve predictability when incorporated into both time-series and artificial intelligence models, illustrating a transferable approach for hydroclimatic prediction that can be extended to water quality forecasting contexts (Kim et al. 2019). Moreover, evidence reviewed in this book indicates that teleconnection-informed analysis can support the selection of the most relevant indices and timescales for a given basin, helping to avoid uncritical index use and to improve model parsimony (Neog et al. 2024). The synthesis is that climate indices should not be added indiscriminately; rather, they should be selected based on mechanistic relevance, seasonality, and the spatial coherence of teleconnection impacts.

5.3.3 Model Evaluation, Uncertainty, and Transferability

Because predictive models can strongly influence decision-making, the study stressed the importance of rigorous evaluation. Environmental modelling is vulnerable to overfitting, spurious correlation, and performance inflation when validation procedures are weak or when predictors indirectly encode information leakage. Classic work on uncertainty in water quality modelling highlighted that uncertainty arises from data limitations, structural choices, and parameterization, and that decision relevance depends on acknowledging these sources rather than reporting point predictions alone (Beck 1987). Similarly, hydrologic model validation requires careful interpretation of goodness-of-fit measures and alignment between evaluation metrics and management objectives, as high numerical performance does not always imply reliable predictive capacity in new conditions (Legates and McCabe 1999).

A further challenge is transferability. Models trained in one basin or under one climatic regime may not generalize under different land-use, hydrogeological, or climate conditions, particularly as non-stationarity increases. The synthesis advanced in this book is that transferability improves when models incorporate physically meaningful predictors such as climate indices, when uncertainty is quantified, and when prediction targets align with operational decisions (e.g., risk classes, threshold exceedance probabilities, or water quality index categories). In this sense, predictive modelling becomes an enabling tool for sustainable water management only when it is embedded within transparent evaluation protocols and governance-ready communication of uncertainty.

5.4 Enhancing Water Resilience with Special Emphasis on Carbon Neutrality

Water resilience has traditionally been defined as a system's capacity to maintain essential functions under shocks such as droughts, floods, and infrastructure failures. This study expanded on this concept by arguing that resilience strategies must now be designed to meet a dual objective: strengthening water security amid increasing climatic variability while advancing carbon neutrality. This reframing is necessary because many conventional water security responses—deep groundwater pumping, long-distance transfers, and advanced treatment—can be highly energy-intensive, thereby risking increased greenhouse gas emissions. The synthesis presented in this book is that resilience must be assessed not only by hydrological performance and water quality protection, but also by the energy and carbon consequences of interventions.

A first principle emphasized here is that climate change and human activities will continue to increase the frequency and intensity of water-related extremes, including compound events. As discussed earlier in this book, such extremes disproportionately shape water availability, degrade quality, and disrupt infrastructure, thereby revealing the importance of buffering capacity and adaptive flexibility (Sillmann et al. 2017; Zscheischler et al. 2020). Resilience, therefore, is not achieved through a single supply augmentation measure, but through portfolios that diversify sources, stabilize demand, and improve system capacity to absorb shocks. The carbon-neutrality lens strengthens this argument: resilience portfolios are preferable because they enable the selection of measures that deliver co-benefits and avoid lock-in to energy-intensive pathways.

5.4.1 Nature-Based and Hybrid Interventions as Low-Carbon Resilience Pathways

The nature-based and hybrid interventions can strengthen water resilience while supporting carbon-neutral objectives. Measures such as rainwater harvesting, runoff management, infiltration enhancement, and filtration-based approaches reduce dependence on high-energy supply solutions by improving local buffering, enhancing recharge, and stabilizing seasonal availability. These measures can also reduce pollutant transport through source control and natural attenuation, thereby improving quality alongside quantity outcomes. In a coupled-systems framework, interventions that modify runoff generation and recharge pathways can reduce the sensitivity of water quality to quantity extremes by stabilizing flow regimes and reducing the magnitude of contaminant pulses during high-flow events.

A key scientific motivation for such strategies is the growing evidence that conventional large-scale models may underestimate observed storage trends and dynamics, implying that local and regional storage management remains critical for both resilience and sustainability. Satellite-derived assessments of terrestrial water storage have highlighted substantial decadal trends that are not fully captured in global models, reinforcing the need for practical recharge, storage, and demand-side resilience measures at basin and sub-basin scales (Scanlon et al. 2018). From a carbon-neutral perspective, interventions that increase effective storage and reduce reliance on pumping and treatment can yield emissions benefits while improving drought robustness.

5.4.2 The Water–Energy–Carbon Nexus in Supply and Treatment Choices

The book's synthesis highlights that water quantity and quality management decisions are embedded within a broader water–energy–carbon nexus. When water quality deteriorates, systems often respond by increasing treatment intensity, seeking alternative sources, or expanding conveyance, all of which can increase energy use. Likewise, when water supply is constrained, systems may adopt energy-intensive solutions, such as deeper pumping or desalination. The carbon-neutrality framing, therefore, requires explicit accounting of energy demand and the consequences of emissions when selecting resilience measures.

A sustainability-oriented approach must evaluate trade-offs between near-term reliability and long-term climate mitigation goals. The synthesis in this book suggests that decarbonization pathways for the water sector should prioritize (i) reducing demand through efficiency and loss control, (ii) increasing local storage and recharge where feasible, and (iii) improving operational optimization using predictive signals and decision-support. This sequencing is important because it can reduce the scale of high-energy interventions required. Although desalination and advanced treatment can be essential in some contexts, the carbon-neutral resilience perspective argues that their role should be optimized and complemented by lower-carbon measures whenever feasible.

5.4.3 Climate-Informed Operation and Adaptive Resilience

Another resilience theme developed is that operational adaptation can deliver substantial benefits, particularly when informed by climate diagnostics and predictive tools. Climate indices and teleconnection signals can provide lead time for managing reservoir operations, groundwater abstraction, and treatment planning, thereby reducing the need for emergency responses, which are typically energy-inefficient and costly. Evidence that climate indices can improve inflow predictability and support forecasting frameworks suggests that climate-informed operations can be an important component of resilience portfolios, especially when combined with artificial intelligence or time-series models (Kim et al. 2019; Agarwal et al. 2024). From a carbon-neutrality perspective, improved forecasting and proactive operations can reduce unnecessary pumping, avoid inefficient treatment escalation, and support better scheduling of energy use.

At the same time, the book emphasizes that operational adaptation must be embedded within uncertainty-aware decision-making. Uncertainty in climate drivers, hydrological response, and quality outcomes can propagate through operational choices, creating risks of overconfidence. Early work on uncertainty in water quality modelling emphasized that uncertainty is intrinsic to environmental prediction and must be managed through transparent assessment rather than ignored (Beck 1987). Similarly, hydrological validation principles remind us that high model fit does not guarantee robust performance under changing conditions, reinforcing the need for adaptive operating rules rather than fixed optimization (Legates and McCabe 1999). The synthesis is that adaptive resilience requires both improved predictive tools and governance mechanisms that can update decisions as new information emerges.

Finally, the study underscored that resilience with carbon-neutral emphasis cannot be achieved through technology alone; it requires enabling governance, financing, and institutional coordination. Groundwater has long been recognized as central to livelihoods and development, but its governance challenges—particularly over-abstraction and quality degradation—remain barriers to sustainable resilience (Shah 2005). The book's integrated perspective suggests that resilience measures must be supported by policies that manage demand, protect recharge zones, regulate pollutant loading, and incentivize low-carbon operational strategies. In practice, this includes strengthening monitoring systems, ensuring accountability for water quality outcomes, and developing decision-support frameworks that integrate quality, quantity, and emissions objectives.

5.5 Roadmap for Next-Generation Water Sustainability

Sustainable water management in the coming decades will depend on treating climate forcing, anthropogenic pressures, water quantity variability, and water quality constraints as parts of a single coupled system. The central forward-looking message of this book is integration: integrating data streams across scales, integrating models that connect climate–hydrology–quality processes, and integrating governance so that planning and operations can respond to compound and non-stationary risks rather than isolated hazards. A priority is closing monitoring gaps by combining satellite-based diagnostics with local observations to track changes in storage and emerging stress. Satellite-derived terrestrial water storage has already shown that large-scale models can underestimate decadal depletion and recovery, reinforcing the value of independent observations for sustainable yield estimation and early warning (Scanlon et al. 2018). The operational challenge is scale: converting coarse signals into actionable basin indicators through workflows that explicitly manage uncertainty and fuse remote sensing with ground networks, particularly where monitoring is sparse (Panday and Kumar 2024). Figure 5.2 illustrates recent state-of-the-art developments in hydrological and water-quality modelling, highlighting emerging analytical paradigms over the last two years. These approaches collectively emphasize stronger physical consistency, scalability across regions, explicit treatment of uncertainty, causal reasoning, and integration of multimodal data streams for decision-oriented forecasting.

Predictive modelling must also evolve from 'best-fit' forecasting toward governance-ready decision support. AI and hybrid machine learning approaches can support anticipatory management for hydroclimatic variables and water quality when validation and predictor selection are rigorous (Abbot and Marohasy 2012; Aldhyani et al. 2020; Bui et al. 2020). The next step is to embed climate information explicitly—using drought indices and climate modes to improve robustness under non-stationarity and extend lead times—while ensuring models remain interpretable and uncertainty-aware (Vicente-Serrano et al. 2010; Kim et al. 2019; Nalley et al. 2019).

Finally, next-generation sustainability requires aligning resilience with carbon neutrality and adaptive governance. Portfolios that reduce demand, enhance recharge and storage, and improve operational efficiency can avoid energy-intensive lock-in

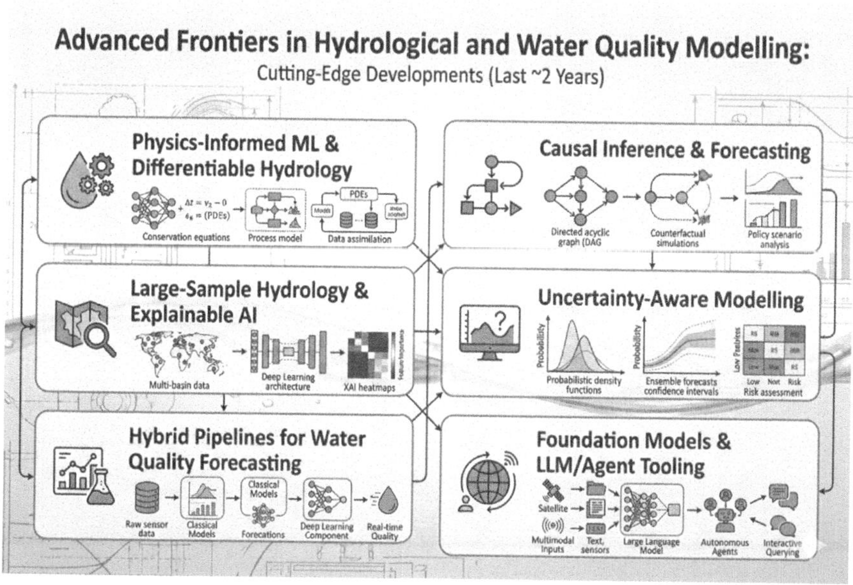

Fig. 5.2 Advanced frontiers in hydrological and water quality modelling. Key trends include physics-informed and uncertainty-aware ML, causal forecasting, large-sample explainable AI, hybrid pipelines, and foundation-model/LLM-based decision support

while strengthening drought and supply robustness (Scanlon et al. 2018). Because uncertainty is inherent and unavoidable, it must be treated as a design objective in monitoring and modelling, not an afterthought (Beck 1987; Legates and McCabe 1999). Delivering these outcomes ultimately depends on institutions capable of integrating evidence into action and protecting groundwater and public health under coupled risks (Shah 2005). Table 5.1 presents a structured synthesis of the conceptual, analytical, and policy insights developed across this study. It integrates climate-driven hydrological transformations, anthropogenic amplification processes, water-quality–quantity interlinkages, advances in predictive modelling, carbon-neutral resilience strategies, and governance reforms into a unified framework. The table identifies critical system interactions, knowledge gaps, strategic research priorities, and implementation pathways, while aligning proposed directions with relevant Sustainable Development Goals (SDGs).

5.6 Conclusions

Water sustainability, as synthesized in this book, depends on recognizing that climate forcing and anthropogenic pressures act together to reshape both water quantity and water quality, often through extremes and non-stationary change. Because

Table 5.1 Integrated synthesis and forward directions for bridging water quality and quantity in the Anthropocene

Thematic domain	Core insights established	Key system interactions	Critical gaps identified	Strategic research directions	Policy and governance priorities	Technological and implementation pathways	Relevant SDGs
Climate-driven hydrological transformation	Climate non-stationarity reshapes precipitation, temperature regimes, recharge, and storage	Flood mobilization versus drought concentration; recharge–evaporation imbalance	Limited regional integration of compound extremes in planning	High-resolution climate–hydrology coupling; compound event modelling	Climate-informed allocation frameworks	Real-time hydro-climatic monitoring; early warning systems	SDG 6, SDG 13
Anthropogenic amplification	Urbanization, agriculture, and industry intensify abstraction and pollutant loads	Groundwater depletion reduces baseflow and dilution	Weak integration of land-use and water planning	Basin-scale abstraction impact assessment	Cross-sector coordination mechanisms	Managed aquifer recharge; green infrastructure	SDG 6, SDG 12
Quality–quantity interlinkages	Dilution–concentration dynamics govern water usability	Storage–contamination feedback loops	Underrepresentation in allocation models	Integrated hydrochemical–hydrological modelling	Quality-integrated water allocation rules	Real-time water quality sensors	SDG 6
Surface–groundwater coupling	Conjunctive management enhances resilience	Baseflow–aquifer connectivity under stress	Separate management of connected systems	Coupled modelling frameworks	Conjunctive governance systems	Recharge enhancement and controlled pumping	SDG 6, SDG 15
Geogenic contaminants	Climate variability alters redox and mobilization patterns	Recharge variability influences arsenic/fluoride dynamics	Limited predictive frameworks	Reactive transport modelling	Groundwater quality zoning	Targeted aquifer remediation	SDG 3, SDG 6
Emerging contaminants	Extreme events alter transport pathways	Flood pulses versus low-flow accumulation	Inadequate removal in conventional treatment	Climate-sensitive contaminant fate modelling	Strengthened effluent standards	Advanced oxidation and membrane systems	SDG 6, SDG 12
Predictive modelling for contaminant forecasting	Data-driven and mechanistic models enhance anticipatory capacity	Integration of land-use, climate, and hydrochemistry	Uncertainty propagation under compound extremes	AI–ML hybrid hydrochemical forecasting	Data-sharing governance frameworks	Digital twins of water systems	SDG 9, SDG 6
Data and monitoring systems	Remote sensing improves spatial resolution	Monitoring gaps limit early detection	Sparse groundwater quality data	Sensor network expansion: satellite–ground integration	National water data repositories	Smart water grids	SDG 9

(continued)

Table 5.1 (continued)

Thematic domain	Core insights established	Key system interactions	Critical gaps identified	Strategic research directions	Policy and governance priorities	Technological and implementation pathways	Relevant SDGs
Carbon-neutral water resilience	Energy–water nexus critical for sustainability	Treatment and pumping energy trade-offs	High carbon intensity of desalination	Low-energy treatment research	Energy-integrated water policy	Renewable-powered desalination; energy-efficient pumping	SDG 7, SDG 13
Nature-based resilience	Wetlands and recharge zones enhance buffering	Ecological restoration improves the quality and storage	Underutilized ecosystem-based solutions	Quantification of ecosystem service valuation	Payment for ecosystem services schemes	Floodplain restoration; green recharge corridors	SDG 15
Infrastructure transformation	Aging systems are vulnerable to extremes	Infrastructure stress amplifies socio-economic risk	Reactive rather than adaptive planning	Climate-resilient infrastructure design metrics	Risk-based infrastructure codes	Decentralized treatment systems	SDG 9, SDG 11
Socio-economic equity	Water insecurity disproportionately affects vulnerable groups	Governance gaps amplify inequality	Limited inclusion in water planning	Equity-based water risk modelling	Inclusive allocation and tariff systems	Community-level monitoring systems	SDG 5, SDG 10
Governance transformation	Fragmented institutions limit adaptive capacity	Policy lag under hydrological non-stationarity	Weak science–policy interface	Adaptive governance experimentation	Integrated IWRM reforms	Decision-support platforms	SDG 16
Integrated resilience pathway	Bridging quality and quantity is essential for long-term sustainability	Multiscale feedback loops across climate–hydrology–society	Siloed disciplinary approaches	Transdisciplinary modelling frameworks	Multi-level water governance reform	Integrated digital + nature-based hybrid systems	SDG 6, SDG 13

usable water is constrained by safety as much as by volume, resilience must be designed around coupled risks, supported by climate-informed diagnostics, predictive modelling, and interventions that stabilize hydrological pathways while reducing exposure and vulnerability. Looking forward, the most promising pathway is integration: integrating satellite and in situ observations to close monitoring gaps, integrating climate signals into forecasting tools for proactive risk management, and integrating carbon-neutral objectives into resilience portfolios to avoid energy-intensive lock-in. Achieving these goals ultimately depends on governance capacity—institutions that can coordinate across sectors, communicate risk and uncertainty, and implement adaptive management that protects public health and long-term sustainability while supporting development needs.

References

Abbot J, Marohasy J (2012) Application of artificial neural networks to rainfall forecasting in Queensland, Australia. Adv Atmos Sci 29(4):717–730. https://doi.org/10.1007/s00376-012-1259-9

Agarwal V, Kumar M, Panday DP, Zang J, Munoz-Arriola F (2024) Unlocking the potential of remote sensing for arsenic contamination detection and management: challenges and perspectives. Curr Opin Environ Sci Health 100578. https://doi.org/10.1016/j.coesh.2024.100578

Aldhyani TH, Al-Yaari M, Alkahtani H, Maashi M (2020) Water quality prediction using artificial intelligence algorithms. Appl Bionics Biomech

Ali Z, Hussain I, Faisal M, Nazir HM, Moemen MAE, Hussain T, Shamsuddin S (2017) A novel multi-scalar drought index for monitoring drought: the standardized precipitation temperature index. Water Resour Manage 31(15):4957–4969

Beck MB (1987) Water quality modeling: a review of the analysis of uncertainty. Water Resour Res 23(8):1393–1442. https://doi.org/10.1029/WR023i008p01393

Bui DT, Khosravi K, Tiefenbacher J, Nguyen H, Kazakis N (2020) Improving prediction of water quality indices using novel hybrid machine-learning algorithms. Sci Total Environ 721:137612

Central Groundwater Board (CGWB) (2018) Groundwater quality in shallow aquifers in India. Central Groundwater Board. http://cgwb.gov.in/WQ/Ground%20Water%20Book-F.pdf

Kim T, Shin JY, Kim H, Kim S, Heo JH (2019) The use of large-scale climate indices in monthly reservoir inflow forecasting and its application on time series and artificial intelligence models. Water (Switzerland) 11(2). https://doi.org/10.3390/w11020374

Legates DR, McCabe GJ Jr (1999) Evaluating the use of "goodness-of-fit" measures in hydrologic and hydroclimatic model validation. Water Resour Res 35(1):233–241

Nalley D, Adamowski J, Biswas A, Gharabaghi B, Hu W (2019) A multiscale and multivariate analysis of precipitation and streamflow variability in relation to ENSO, NAO, and PDO. J Hydrol 574:288–307. https://doi.org/10.1016/J.JHYDROL.2019.04.024

Neog N, Goswami R, Panday DP, Kumar A, Selvan MT, Baruah A, Kumar M (2024) Arsenic contamination in the groundwater of Northeastern India: critical understandings on geotectonic controls and the need for intervention. Curr Opin Environ Sci Health 100539. https://doi.org/10.1016/j.coesh.2024.100539

Panday DP, Kumar M, Agarwal V, Torres-Martínez JA, Mahlknecht J (2024) Corroboration of arsenic variation over the Indian Peninsula through standardized precipitation evapotranspiration indices and groundwater level fluctuations: water quantity indicators for water quality prediction. Sci Total Environ 176339. https://doi.org/10.1016/j.scitotenv.2024.176339

Panday DP, Kumar M (2024) Decoding uranium variation over the Indian Peninsula through leveraging standardized precipitation evapotranspiration indices and groundwater level fluctuations. Groundw Sustain Dev 101241. https://doi.org/10.1016/J.GSD.2024.101241

Scanlon BR, Zhang Z, Save H, Sun AY, Schmied HM, Van Beek LPH, Wiese DN, Wada Y, Long D, Reedy RC, Longuevergne L, Döll P, Bierkens MFP (2018) Global models underestimate large decadal declining and rising water storage trends relative to GRACE satellite data. Proc Natl Acad Sci USA 115(6):E1080–E1089. https://doi.org/10.1073/PNAS.1704665115

Shah T (2005) Groundwater and human development: challenges and opportunities in livelihoods and environment. Water Sci Technol 51(8):27–37

Shah D, Shah HL, Dave HM, Mishra V (2021) Contrasting influence of human activities on agricultural and hydrological droughts in India. Sci Total Environ 774:144959

Sillmann J, Thorarinsdottir T, Keenlyside N, Schaller N, Alexander LV, Hegerl G, Seneviratne SI, Vautard R, Zhang X, Zwiers FW (2017) Understanding, modelling and predicting weather and climate extremes: challenges and opportunities. Weather Clim Extremes 18:65–74. https://doi.org/10.1016/J.WACE.2017.10.003

Tripathi S, Panday DP, Kumar M (2025) Emerging contaminants and water conservation: challenges, strategies, and solutions for sustainable water management. Water Scarcity Manage: Enabling Technol 1–17

Vengosh A, Coyte RM, Podgorski J, Johnson TM (2022) A critical review on the occurrence and distribution of the uranium-and thorium-decay nuclides and their effect on the quality of groundwater. Sci Total Environ 808:151914

Vicente-Serrano SM, Beguería S, López-Moreno JI (2010) A multi-scalar drought index sensitive to global warming: the standardized precipitation evapotranspiration index. J Clim 23(7):1696–1718

World Health Organization (2004) Guidelines for drinking-water quality, vol 1. World Health Organization

Zscheischler J, Martius O, Westra S, Bevacqua E, Raymond C, Horton RM et al (2020) A typology of compound weather and climate events. Nat Rev Earth Environ 1(7):333–347

The manufacturer's authorised representative in the EU is Springer Nature Customer Service Centre GmbH, Europaplatz 3, 69115 Heidelberg, Germany. If you have any concerns regarding our products, please contact ProductSafety@springernature.com

Printed and bound by CPI Group (UK) Ltd, Croydon, CR0 4YY
21/08/2026
02202063-0001